About CIB and the CIB series

CIB, the International Council for Research and Innovation in Building and Construction, was established in 1953 to stimulate and facilitate international cooperation and information exchange between governmental research institutes in the building and construction sector, with an emphasis on those institutes engaged in technical fields of research.

CIB has since developed into a worldwide network of over 5,000 experts from about 500 member organisations active in the research community, in industry or in education, who cooperate and exchange information in over 50 CIB Commissions and Task Groups covering all fields in building and construction related research and innovation.

http://www.cibworld.nl/

This series consists of a careful selection of state-of-the-art reports and conference proceedings from CIB activities.

Culture in International Construction W. Tijhuis et al.
ISBN: 9780415472753. Published 2012

R&D Investment and Impact in the Global Construction Industry
K. Hampson et al.
ISBN: 9780415859134. Published 2014

Public Private Partnership A. Akintoye et al.
ISBN: 9780415728966. Published 2015

Court-Connected Construction Mediation Practice A. Agapiou and D. Ilter
ISBN: 9781138810105. Published 2016

Advances in Construction ICT and e-Business S. Perera et al.
ISBN: 9781138914582. Published 2017

Clients and Users in Construction K. Haugbølle and D. Boyd
ISBN: 9781138786868. Published 2017

Clients and Users in Construction

Clients have been identified as critical for building delivery but have been under-researched with only a few studies about them. This book seeks to address this gap.

A deeper look into the nature of construction clients and their relation to building users exposes more fundamental questions related to the activity of building and the activity in the building. These fundamental questions include 'How do clients get what they want?', 'How do clients cope with the building process?', and 'How are clients being shaped by building(s)?'.

This book on clients and users is structured around three main themes:

- Agency is concerned with the classical agency/structure dichotomy on actions, roles and responsibilities or, put differently, whether actors can act freely or are bound by structural constraints.
- Governance is related to the interplay between clients and the supply system: clients govern the supply system but are at the same time governed by the supply system through different processes and mechanisms.
- Innovation deals with change versus stability, and what part clients and users play in this struggle.

The book includes theoretical and conceptual frameworks on what constitutes clients and users as well as case studies on R&D themes of relevance to practice.

Kim Haugbølle conducts advisory services to the Danish government, undertakes teaching, and develops research-based knowledge to improve the built environment. He has authored or co-authored more than 200 publications on innovation, procurement, life cycle economics and sustainable design with a special emphasis on the role of clients. He has been involved in the coordination and management of several national and international R&D projects, and has been heading the secretariat of a think tank and a research department. He is the international co-coordinator of the CIB Working Commission W118 on Clients and Users in Construction as well as a board member of the Nordic researchers' network on construction economics and organisation (CREON).

David Boyd is Professor of Construction at Birmingham City University, UK. He has a background in engineering, but is better known for his management insights of the industry. His major contribution has been to develop a model of projects in the industry as complex adaptive socio-technical systems. His earlier research on construction clients has been published in the book *Understanding the Construction Client* which was adopted by the UK Construction Clients Forum. He is developing research into practice through the philosophy of expertise-in-context and is currently researching the challenges of connecting human and information perspectives in BIM. He is the international co-coordinator of the CIB Working Commission W118 on Clients and Users in Construction.

Clients and Users in Construction

Agency, Governance and Innovation

Edited by
Kim Haugbølle and David Boyd

Routledge
Taylor & Francis Group

LONDON AND NEW YORK

First published 2017 by Routledge

2 Park Square, Milton Park, Abingdon, Oxon, OX14 4RN
605 Third Avenue, New York, NY 10017

Routledge is an imprint of the Taylor & Francis Group, an informa business

First issued in paperback 2020

British Library Cataloguing-in-Publication Data
A catalogue record for this book is available from the British Library

Library of Congress Cataloging in Publication Data
A catalog record for this book has been requested

ISBN: 978-1-138-78686-8 (hbk)
ISBN: 978-0-367-73616-3 (pbk)

Typeset in Goudy
by Saxon Graphics Ltd, Derby

Contents

Figures

Tables

Preface

CIB, the International Council for Research and Innovation in Building and Construction, supports international collaboration and knowledge exchange through its 50 permanent Working Commissions and temporary Task Groups. In 2010, the Board of CIB decided to establish a new Working Commission W118 on Clients and Users in Construction. This decision followed after a three-year period 2007-10 in which CIB explored the possibility and potential of establishing a fourth Priority Theme on 'Clients and Users' as part of its new proactive approach.

The aim of this Working Commission is to:

- bring together the experience and expertise of researchers and practitioners who would otherwise not have interacted with each other within both CIB and other relevant associations, organisations and networks;
- develop, share and disseminate appropriate research methodologies and theories as well as practices with regard to the successful client management of procurement, innovation, and agency of users; and,
- encourage and enable new collaborative, multi-disciplinary research activities to take place through the establishment of a critical mass of interested and diverse researchers and practitioners both within and outside CIB.

As coordinators of the working commission, we are proud to present this book as one of the central outcomes of the activities of the commission.

The primary readership of this book is expected to be academics and students who share an interest in a much closer view at how clients and users behave and what is being valued by them. These academics and students may work or study within:

- architectural or design management,
- procurement systems,
- construction management, and
- real estate, property management and facility management.

A new readership in business schools is also emerging among business and management students, particularly MBAs.

Some professional practitioners may also purchase the book to improve their reflexive practices and to stimulate their own concerns and ambitions to develop new strategies, policies and actions. The most likely professional readership may include the following groups:

- All organisational levels of construction client organisations, senior consultants at architectural or engineering offices, and construction managers at contractors.
- Advisors in trade associations.
- Policy makers at national ministries, international policy makers at, for example, the European Commission and United Nations, senior level advisors at local/regional authorities, and funding officers at research funding agencies.

Professional readers and academics may find an interest in the book to get a better grasp on how to create public policies, supporting mechanisms and regulation that can enhance a competitive construction industry, provide better value for customers and stimulate innovation.

As editors of this book we are very pleased to recognise the diversity of contributors and contributions made for this book. They span theoretical and conceptual frameworks on what constitutes clients and users as well as case studies on research and development themes of relevance to practice. This book presents an international outlook with contributions from Australia, Hong Kong/PRC, South Korea, Nigeria, Sweden, Denmark, The Netherlands, the United Kingdom, Northern Ireland and France. The contributors include both members and non-members of the working commission, and researchers as well as practitioners.

As coordinators of the working commission CIB W118 we look forward to continue working with the contributors on improving our understanding of the role of clients and users in construction.

Kim Haugbølle
David Boyd

Foreword by CIB

Wim Bakens, Secretary General

In many countries the building and construction industry in recent years concluded that it might be necessary for medium or long-term survival to redefine itself.

Platforms of challenges and opportunities for such industry redefinition may be found in:

- Technology: including ICT, internet, BIM and the like as well as concepts such as offsite manufacturing and the need for technological innovation in general.
- Society: economic, social, geopolitical and demographic developments and expectations, including the need for sustainability and resilience.
- Process: changing organisational and management concepts, the need for integral procurement and a fast changing insurance and legal context.

In whatever direction developments may lead and how this may impact upon the building and construction industry, in most Western countries no attempts towards an industry re-definition has a chance for success, unless one crucial condition is met. New approaches and new concepts have to take account of well-informed and well-resourced clients and users for the industry's deliverables, products and services. Their wishes and expectations need to be the focus of attention for the development of new practices in the industry, as this has been the case in other successful, innovative and future-oriented industries.

As this book shows, clients of the building and construction industry and users of buildings and the built environment as designed, engineered, constructed and delivered by this industry are becoming better informed, better organised, better articulated and, in general, more professional. Taking them very seriously and investing permanently in trying to gain their trust are indeed crucial conditions for the industry's survival.

I wholeheartedly recommend this book to all clients for and users of buildings and the built environment and to the industry that services them.

Acknowledgements

The editors wish to thank all those who have made this publication possible through their encouragement, contributions and support.

We first wish to thank the international group of authors who have contributed to the chapters of this book. The authors have tirelessly tried to accommodate the commentaries provided on earlier drafts of their contributions in order to bring their work to the highest possible level. We highly appreciate their efforts.

This publication is the outcome of the International Council for Research and Innovation in Building and Construction's (CIB) ambition to strengthen its efforts towards understanding the role of clients and users in construction. As a result CIB established a permanent Working Commission CIB W118 on Clients and Users in Construction. We therefore extend our sincere thanks to the CIB Board and CIB Secretariat who have both facilitated and encouraged the establishment of the commission and this book's endeavour.

The editors wish to thank the editorial staff at Routledge/Taylor & Francis. In particular, we wish to thank editorial assistants Brian Guerin and Matthew Turpie and not least senior editorial assistant Alice Aldous for giving us this unique opportunity to produce this book. We would also like to expressly thank those anonymous reviewers who provided external review of this book's intent and results.

The editors also wish to thank their colleagues and home universities for encouragement and financial support. Without this support this publication would not have been realised.

Finally, the editors would like to acknowledge those who have granted permission to reproduce their material in this book.

Contributors

Solomon Adjei, Lecturer in Quantity Surveying, School of the Built Environment, Faculty of Science Engineering and Built Environment (CEBE), Birmingham City University. Solomon is a PhD holder from the University of Wolverhampton, where he researched waste management practices in the UK construction industry. He completed his first degree in building technology from the Kwame Nkrumah University of Science and Technology, Ghana, in 2009, after which he worked as a teaching and research assistant at the same university for over a year. Between 2009 and 2011 he also practiced as a quantity surveyor and a project manager in Ghana until he joined the University of Wolverhampton for his PhD. His research interests include: sustainable construction, sustainability and construction law, innovation in the construction industry and contract practice.

Abdul-Rasheed Amidu, Senior Lecturer, University of Auckland, New Zealand. Abdul-Rasheed is senior lecturer in property with focus on property finance, investment and investment decision making. Although Abdul-Rasheed's background is primarily in property with research focused on finance, investment and valuation decision making, he has for many years taught construction management-related courses including business management, which explores the strategic and operational issues in the construction context. Abdul-Rasheed is a professional member of the Royal Institution of Charted Surveyors (MRICS).

Henrik Lindved Bang, Director of the Danish Association of Construction Clients (Bygherreforeningen). Henrik has a background in civil engineering and construction management. He has previously been active in academia at Copenhagen Business School (CBS) and the Danish Building Research Institute (SBi) conducting research on strategies, economics and organisation of construction activity. Since 2004, he has been heading the secretariat of the Danish Association of Construction Clients (Bygherreforeningen), one of the leading trade associations in the Danish construction sector.

Niels Haldor Bertelsen, Senior Researcher, Danish Building Research Institute, Aalborg University. Niels is doing research on construction management and innovation. His research background is in building physics, moisture transport, indoor climate, and building and insulation materials. His latest research is on productivity, innovation, process control and education in close cooperation with authorities, building owners, advisers, constructors, suppliers and teachers. Over the past thirty years he has established several innovative networks with different sector partners, e.g. on insulation materials, indoor climate and inspection of housing, and most recently the Danish network AlmenNet with social housing clients.

Frédéric Bougrain, Researcher, Centre Scientifique et Technique du Bâtiment (CSTB), Champs-sur-Marne, France. Frédéric works as a researcher for CSTB (a state-owned industrial and commercial research centre under the wings of the Ministry of Housing) in France. His research interests range across construction economics and management, with a particular focus on innovation, energy saving performance contracts, regulatory impact analysis and more recently building information models. He previously lectured at the University of Orléans (France), where he defended his thesis on innovation, small and medium-sized enterprises and the consequences for regional technology policy. Frédéric has published papers on public–private partnerships, energy saving performance contracts, innovation in small and medium-sized enterprises and the social housing sector.

David Boyd, Professor of Construction and Director, Centre for Environment and Society Research, Birmingham City University, UK. David has a background in engineering, but is better known for his management insights of the industry. His major contribution has been to develop a model of projects in the industry as complex adaptive socio-technical systems. He has completed research on construction clients which was published in the book *Understanding the Construction Client* co-authored with Ezekiel Chinyio and adopted by the Construction Clients Forum in UK. He is developing research into practice through the philosophy of expertise-in-context and is currently researching the challenges of connecting human and information perspectives in BIM. He is the international co-coordinator with Kim Haugbølle of the CIB Working Commission W118 on Clients and Users in Construction.

Ezekiel Chinyio, Senior Lecturer, School of Architecture and Built Environment, University of Wolverhampton. Ezekiel has an interest in diverse construction management aspects such as clients' requirements, briefing, stakeholder management, risk management, contractor selection, contracts, and pedagogy. His current research is examining the impacts of psychology on the built environment. This work is exploring the engineering and delivery of a built environment that will provide full satisfaction to humanity. Ezekiel has published widely and is a member of the CIOB and some CIB task groups. He reviews articles for many leading construction journals and books for international publishers.

Paula Femenías, Associate Professor, Department of Architecture, Chalmers University of Technology, Sweden. Femenías works mainly in inter- and transdisciplinary environments projects with a broad approach to analysis and support to progress innovation for more sustainable building, including qualitative, quantitative and artistic research. A specific research interest is renewal and transformation of existing stocks. Her work includes papers developing the perspective of clients and users, but also multi-stakeholder and multi-value models and architectural knowledge.

Ada Fung, Deputy Director of Housing (Development and Construction), Hong Kong Housing Authority, Hong Kong/PRC. She is an active member in the architectural field as well as in the construction industry in Hong Kong. As the Deputy Director of Housing she supervises the Development and Construction Division of the Housing Department, overseeing all facets of public housing development and construction work in Hong Kong, from planning, design, contract management to completion, as well as establishing operational policies on procurement, quality, performance assessment, dispute resolution, research and development, safety and environmental issues. She also promotes partnering, value management, risk management, ethical integrity, site safety, corporate social responsibility, sustainable development, community engagement, green building, Building Information Modelling and product certification in the industry.

Stefan Christoffer Gottlieb, Senior Researcher, Danish Building Research Institute, Aalborg University. Stefan is engaged in a wide range of activities, including public sector consultancy, European research projects and university teaching. His primary research and teaching are in the area of institutional change and the impact of regulation on working practices. He is currently engaged in research on the consequences of the harmonisation of European markets for the Danish construction sector, with emphasis on the dynamics between innovation and competitiveness on one hand, and considerations to local building customs and the quality of the built solutions on the other.

Kim Haugbølle, Senior Researcher, Danish Building Research Institute (SBi), Aalborg University, Copenhagen, Denmark. Kim Haugbølle conducts advisory services to the Danish government, undertakes teaching, provides training of professionals and develops research-based knowledge to improve the built environment. He has authored or co-authored more than 200 publications on innovation and socio-technical change in the construction industry with a special focus on the role of the construction client, life cycle economics and sustainability. He has headed the secretariat of the think tank Danish Building Development Council and later a research department at the national building research institute. Kim is the international co-coordinator of the CIB Working Commission W118 on Clients and Users in Construction as well as a board member of the Nordic researchers' network on construction economics and organisation (CREON).

Marleen Hermans, Professor at Delft University of Technology, Management in the Built Environment. She holds a chair on public commissioning in construction. Both in research as well as in education she focuses on the role of public construction clients and their leading role in changing the sector by means of procurement. Her chair was established by the Dutch construction clients' forum, and Marleen is the adviser of the steering group. She is also managing partner of Brink Management & Advies, a Dutch consultancy firm in project management, real estate and infrastructure consultancy.

Christopher Heywood, Associate Professor in Property and Management, Faculty of Architecture, Building and Planning, University of Melbourne, Australia. Chris has a professional background in architecture but since 1999 has been developing corporate real estate management as an academic field in Australia. This started with Russell Kenley in the Corporate Real Estate and Asset Management Research Group at the University of Melbourne. Since that project finished Chris has assumed leadership of that development and now works with colleagues around the world to advance the field. He brings a trans-disciplinary perspective to understanding the built environment, its creation, its purposes and its management to achieve those purposes. Among other projects he is currently investigating how organisations align their real estate and business strategies.

Mieke Hoezen, Policy Adviser at the Ministry of Infrastructure, the Netherlands. Mieke Hoezen advises the Highways Agency on organisational knowledge, management, and the quality of the infrastructure networks. She has had several roles within the ministry advising policy makers and public purchasing servants on how to implement lessons learned into new procedures and contracts. Mieke's PhD thesis about the development of contract and contact under influence of the competitive dialogue procedure was awarded with the European Court of Auditors award for audit-related research in 2012. Since then, Mieke has been active in the dissemination of knowledge on public commissioning in her role as Secretary of the Task Group Procurement in the Conference of European Directors of Roads and within the European Academy for Taxes, Economics and Law.

Youngsoo Jung, Professor in Construction Engineering and Informatics, Myongji University, South Korea. Youngsoo Jung holds a PhD in Construction Engineering and Project Management from the University of Texas, Austin. Prior to joining the faculty at Myongji University in 2000, he had eleven years of industry experience as a project engineer, cost engineer and information systems manager. He has over 200 publications including journal articles, proceedings, books and reports. His areas of research interest include project delivery systems (PDS), computer integrated construction (CIC/BIM), and automated cost and time management. Recent efforts focus on the automated lifecycle information management for power plants and capital airports. He is a current vice president of Korea Institute of Construction Engineering and Management (KICEM) and served as the conference chair for the 6th

International Conference on Construction Engineering and Project Management (ICCEPM 2015).

Seunghee Kang, Research Professor, College of Architecture, Myongji University, South Korea. Seunghee Kang had five years of experience as a researcher at public enterprises before he joined Myongji University in 2014. His research interest areas include construction policy, project delivery systems and automated cost and time management. He has conducted several R&D projects funded by the Korean government such as industrialisation policy for modernised Korean housing, supportive policy for super tall buildings and advanced project delivery system. Currently, he is actively involved in a research project which is developing the knowledge-based preliminary critical path method (CPM) scheduling model for airport construction projects from the client or PMO perspective.

Russell Kenley, Professor of Management, Faculty of Business and Law, Swinburne University of Technology, Melbourne, Australia. Russell has a professional background in building, but has over 25 years of research experience in the construction and management of the built environment. He has focussed on the strategic management of corporate real estate and collaboration with Chris Heywood and is now developing new models for project planning and control in both vertical and horizontal infrastructure. Russell has published two books, *Financing Construction: Cash Flows and Cash Farming*, and *Location-based Management for Construction: Planning, Scheduling and Control*. His research develops new understandings of the relationship between clients and the supply chain, and proposes new project management models for improved productivity and governance of projects.

Ciaran McAleenan, Lecturer at Ulster University, Northern Ireland. Ciaran, a chartered civil engineer working in industry since 1978, designed and constructed water industry infrastructure projects, developed health and safety (H&S) and road design standards, and managed major road construction projects before joining Ulster University in 2010, where he co-founded the Civic Ecology and Life Infrastructure (CELI) research group. A member of the Institution of Civil Engineers' editorial board for the *Management, Procurement and Law* journal he sits on its H&S Register Steering Group. A member of CIBW099 and W120, he regularly presents at world and international congresses, and in 2015 he hosted the CIBW099 conference featuring the world's best H&S researchers.

Philip McAleenan, Researcher and OSH Management Consultant, Expert Ease International, Northern Ireland. Philip is co-developer of the 'Operation Analysis and Control' model for safety. His research into workplace culture, leadership and ethics reasoning led to developing the 'Organisation Cultural Maturity Index' and to undergraduate modules on ethics reasoning in construction. He publishes extensively, contributing to safety books in the UK and the USA and to books on workplace culture and leadership. Internationally, he regularly presents to occupational health and safety conferences, including

ILO World Congresses and to professional development in USA and Canada. He contributes to CIB W099 and assisted in the organisation and management of its 2015 conference, titled 'Benefitting Workers and Society through Inherently Safe[r] Construction'.

Karen Mogendorff, previously at Delft University of Technology. Karen is an anthropologist and communication scientist. She conducted interpretive qualitative research at Delft University of Technology, Management in the Built Environment, on how public construction clients conceptualise and translate traditional and new forms of public commissioning in situated organisational contexts. In general, her research focuses on how people or organisations deal with tensions and dilemmas in multi-party settings or centres on how professional and practical expertise may be optimally combined in these settings.

Chimene Obunwo, PhD in Engineering Management from the University of Wolverhampton. Chimene is an astute researcher with a focus on the quality and satisfaction of engineering construction projects within the built environment in developing countries. Currently, his research interests are on quality management in government road construction projects and the optimisation of project management activities that enhance stakeholder satisfaction within the built environment. Chimene is a member of the Association of Researchers in Construction Management (ARCOM).

Rolf Simonsen, Head of projects, Danish Association of Construction Clients (Bygherreforeningen). Rolf has a background in civil engineering and construction management. He has worked academically with lean construction, management concepts and change management at the Danish Technological Institute and is an external lecturer at the Technical University of Denmark. Since 2008 he has been heading the Danish development programme 'The Value Creating Construction Process'. The programme develops new tools and guidelines for the construction sector.

Subashini Suresh, Reader of Construction Project Management, School of Architecture and Built Environment, University of Wolverhampton. She is the Athena Swan Champion for the School. She has over 15 years of teaching, consulting and research experience in a wide range of construction management subject areas in developed and developing countries. Her key areas of interest are: construction project management, knowledge management, building information modelling, health and safety, sustainability/green construction, emerging technologies, quality management, leadership in change management initiatives, organisational competitiveness, business process improvement, lean construction, Six Sigma leadership and soft computing.

Niraj Thurairajah, Senior Lecturer, Birmingham City University. Niraj has a background in building economics, previously holding posts at the University of Moratuwa, Sri Lanka, and University of Salford, UK. His research includes work on education in construction, construction innovation and building information modelling (BIM). His current research focus is on commercial management aspects of BIM and the cultural transformation of construction cost consultants.

Peter Vogelius, Senior Researcher, Danish Building Research Institute, Aalborg University. Peter works with different research themes related to construction management. This includes innovation and collaboration in general. His most recent project is about renovation processes in housing with special focus on indoor climate and user/technology interfaces. Peter also works on the use of ICT in construction, where he has had a special focus on competence building and classification issues. Peter is teaching on the MSc programme on 'Construction Management and Informatics'.

Leentje Volker, Associate Professor in Public Commissioning, Delft University of Technology; and Secretary of the Dutch Construction Client Forum. Specialised in public commissioning in the fields of architecture and infrastructure, Leentje is intrigued by the interaction between people and the built environment on individual, organisational and institutional level. Key topics in her work include partner selection, project governance and strategic asset management. Working closely with the Dutch Construction Client Forum, a joint initiative of public commissioning clients in the Netherlands aimed at improving the professional level of Dutch construction clients, enables her to combine science and practice in her daily work.

Kristian Widén, Associate Professor in Innovation Sciences with focus on the Built Environment, and Research Leader for Sustainable Innovation Management in Building, Halmstad University. Kristian has for many years been conducting research on innovation and innovation diffusion in and for the built environment. Currently the research interests are on innovation and innovation diffusion in the interface between suppliers and contractors, the effect of context and the role of digitisation for innovation and innovation diffusion in the built environment. Kristian has been active member in several CIB task groups addressing innovation from various perspectives and is a board member of the Nordic researchers' network on construction economics and organisation (CREON).

Ka-man Yeung, Hong Kong Housing Authority, Hong Kong/PRC. K. M. Yeung is a Quantity Surveyor by profession. He has over 34 years of experience in the construction field in Hong Kong. He is the head of the Quantity Surveying Section of the Development and Construction Division of the Housing Department overseeing the provision of quantity surveying services and cost control of the construction cost budgets of public housing developments. He has a broad range of experience in procurement arrangements and has actively participated in the development of various contractual documentations and the dispute avoidance and resolution adviser system for mitigating and resolving contract disputes.

Abbreviations

AEC	Architectural, Engineering and Construction industry
BIM	Building Information Model
CAR	Constructor's All Risk
CCG	Construction Clients Group (in both United Kingdom and New Zealand)
CIB	International Council for Research and Innovation in Building and Construction
CM	Construction Management
DACC	Danish Association of Construction Clients (in Danish: Bygherreforeningen)
DCCF	Dutch Construction Clients Forum (in Dutch: het Opdrachtgeversforum)
DK	Denmark
EU	European Union
GBP	Great Britain Pounds
GDP	Gross Domestic Product
H&S	Health & Safety
HKD	Hong Kong Dollars
IDI	Inherent Defects Insurance
KPI	Key Performance Indicators
NL	The Netherlands
NZ	New Zealand
PCP	Pre-commercial Procurement
PDS	Project Delivery System
PI	Professional Indemnity
PMO	Project Management Organisation
R&D	Research and Development
SCC	Swedish Construction Clients (in Swedish: Byggherrarna)
SE	Sweden
TPL	Third Party Liabilities
UK	United Kingdom
US	United States of America
USD	US Dollars
W118	CIB Working Commission W118 on Clients and Users in Construction

Introduction
Three research themes

Kim Haugbølle and David Boyd

Introduction

Construction clients have frequently been called upon to become change agents of the construction industry. These calls have been made by a multitude of actors and through various means like governmental policies, business initiatives and research projects. In recent years, the issue of using (public) procurement as a driver of innovation has also gained increased international interest, for example through the European Commission's Lead Market Initiative and various other international organisations such as the United Nations and ICLEI (International Council for Local Environmental Initiatives).

The overarching raison d'être behind these calls is a wish to redirect attention from the hitherto dominant focus on the production of built facilities to the demand side. Clients and users play a significant role in shaping construction and real estate through various political, economic, social, technological, environmental and legislative drivers. Getting a better grasp of the aspirations, needs and behaviour of users and clients may offer an important new road for the industry to deliver more value for money. Given the political attention towards clients and users, the time seems ripe to address this emerging field of research, development, training and political action in a more concerted manner. Hence, this book will provide:

- an *international outlook* on the challenges facing construction clients and their strategies to cope with these;
- a unique insight into the *fundamentals of the building process* guiding both construction clients and the supplying network;
- exemplars and case studies of *best practice* among construction clients; and
- a theoretically informed *understanding* of the potential and limitations of client-driven change in construction.

A deeper look into the nature of construction clients and their relation to building users exposes more fundamental questions related to the activity *of* building and the activity *in* buildings. These fundamental issues include questions such as 'how clients get what they want', 'how clients cope with the building process' and 'how clients are shaping and being shaped by building(s)'.

The research roadmap of the relatively new CIB Working Commission W118 on Clients and Users in Construction (Haugbølle and Boyd, 2013) has identified three main themes for research and development and this book on clients and users is consequently structured around these three themes:

- Agency is concerned with the classical agency/structure dichotomy on actions, roles and responsibilities, or, put differently, how actors can act freely or are bounded by structural constraints.
- Governance is related to the interplay between clients and the supply system: clients govern the supply system but at the same time they are governed by the supply system through different processes and mechanisms.
- Innovation deals with socio-technical development and what part clients and users play in this struggle between change and stability.

This book is therefore divided into three parts, with each theme addressed in turn. The book includes theoretical and conceptual frameworks on what constitutes clients and users as well as case studies on R&D themes of relevance to practice and academia.

Part I: Agency – roles and responsibilities

As pointed out by the CIB W118 Research Roadmap (Haugbølle and Boyd, 2013), the major problem with regard to the roles and responsibilities of clients and users involves the issue of whether clients and users can act independently to achieve their aim, or whether they are always required to act in the way their environment expects them to act. The significance of this for practice is about how clients and users operate and how they are able to change. In academic terms this is associated with the classical debate on the relationship between agency and structure. Put differently, do socio-technical structures determine the behaviour of actors, or are socio-technical structures the result of human agency? Over time, a range of different positions have evolved in for example sociology, political science and philosophy. One main position is the structuralist perspective in which the agency of actors can largely be explained by reference to the socio-technical structures which more or less determine what actors can do. At the opposite end of the scale, another position underlines the capacity of individual actors to determine the outcome of their actions. In between these two positions, a number of alternative perspectives such as constructivism try to find a more balanced position between the two.

In a subsequent summary of the CIB W118 Research Roadmap, Haugbølle and Boyd (2016) describe three central research objectives within this theme, the first of which is related to the needs and values of clients and users. A core challenge is the need for a shift in focus from building as an end in itself to building as a means to achieve objectives related to the activities of the users of a building during its lifecycle. Consequently, a prerequisite for clients and users is to be knowledgeable about their own values and needs, and to juxtapose, converge or otherwise position these in relation to other stakeholders. Thus, it is important to explore in detail client needs and user values in different domains by mapping the content and scale of clients' and users' value chains in various national and institutional contexts.

The second research objective is related to how the roles and responsibilities of clients and users in construction are shaped by socio-technical structures, and what the implications are. The ability of clients and users to influence chains of actions is not only a result of their own will and wishes, but is shaped in part by how they are embedded in structures that distribute roles and responsibilities to different actors. Public regulation, policy making, market conditions, etc., are some of the framework conditions that influence the practices and behaviour of clients and users. A better understanding of how clients' and users' independent actions are exercised under different structural conditions is essential for identifying the space for action available to clients and users.

The third research challenge is related to the question of what constitute the two terms 'client' and 'user'. An improved understanding of both actor terms and their various configurations under different structural conditions are imperative in order to develop theories and conceptual frameworks which can be used to build a coherent model or even theory of clients and users. Viewed from a management perspective, the problem could be expressed as contingencies, convergence and contradictions in an organisation between the different roles and associated responsibilities. This will be essential for identifying the clients' and users' core competences and capabilities, and how these can be developed further.

Part I of this book addresses issues related to the agency of clients and users. This part consists of four contributions that cover:

- the role and merits of construction clients' forum;
- a conceptual model of clients and users;
- the changing roles of clients and users due to new requirements; and
- the moral and ethical requirements of clients specifically in relation to the occupational health and safety of construction workers as a group of 'users'.

The first contribution related to this theme is delivered by Henrik L. Bang, Marleen Hermans, Rolf Simonsen and Karen Mogendorff, who as representatives of client organisations in Denmark and the Netherlands investigate the relatively recent phenomenon of construction client associations in order to understand the merits of such organisations. A national client association is normally driven by construction clients' shared visions and their sense of responsibility for reforming and professionalising the sector through 'demand pull', cooperation and sharing of knowledge and experiences. An overview of different priorities and inherent strategies is established by comparing characteristics and initiatives of construction client associations across five countries. Drivers that are instrumental to the success of these client associations are discussed including perspectives for future development and the potential spread of such organisations to other countries.

The second contribution, by Chris Heywood and Russell Kenley, uses corporate real estate (CRE) theory to develop a complex model of CRE organisations as space-using clients (by being one) and users (by having them). Within an overarching, demand–supply framework the model assembles sub-models of CRE's organisational roles, CRE's level of strategic-ness and levels of strategic CRE management (CREM) practice and evolution representing internal organisational dynamics of entities procuring construction-related products and services. These dynamics manifest in specific procurement practices with client-ness and user-ness emerging from that dynamism. While focused on commercial and institutional property, the model represents procurement relationships generally.

The third contribution is by Frédéric Bougrain and Paula Femenías and addresses the implications of changing roles of clients and users with regard to the implementation of low-energy buildings, the performance of which relies not only on the quality of design, construction and operation but also on users' behaviour. To illustrate this situation, the study compares cases of low-energy buildings in France and Sweden, countries with similar ambitions but following different policies to implement thermal regulations and evolving within distinctive contexts. The research indicates that there is a mismatch between predicted energy value and the real performance of the building. To limit this gap, there is a need to develop new relationships between clients, occupants and operators and to promote professional clients who can interpret users' needs without jeopardising the energy performance of the building.

The fourth and final contribution within this theme, by Ciaran McAleenan and Philip McAleenan, turns our attention to culture, a concept within which humankind stands as the embodiment and maker of culture and morality, and presupposes agency and the capacity to act independently. Clients simultaneously require organisations they engage to act freely, whilst bound by the structural constraints embodied in their contract. Clients need continuing reassurance that their contractors are competent to act in an ethical and culturally mature manner. This chapter, based on critical theory, develops ethics reasoning into a framework linked to the culture and maturity levels of individuals and organisations illustrating the efficacy of developing these within organisations while delivering a positive effect on meeting clients' requirements.

Part II: Governance – processes and mechanisms

As suggested by the CIB W118 Research Roadmap (Haugbølle and Boyd, 2013, 2016), the second research and development theme to be addressed is governance, meaning the act of governing, which includes the strategies, processes and mechanisms, rules, behaviour, etc., that affect the way power is exercised among actors in and around building and construction. Governance is a multifaceted concept which may take place at different levels: project level, corporate level, or regulatory level. This last is particular relevant for public clients and in relation to urban and infrastructure developments. Good governance is often associated with transparency, accountability, participation, etc.

The research roadmap (Haugbølle and Boyd, 2016: 5) goes on to suggest three research objectives to be addressed under the governance theme:

- analysing clients' and users' strategies, competences, capabilities and practices for procurement, management and use of buildings and constructions in a life-cycle perspective;
- evaluating participatory methods and tools to involve users and stakeholders in decision-making processes on construction as well as operation; and
- understanding the mechanisms behind successful/failed projects and why some tools, procurement routes, etc., may be more appropriate than others.

The first challenge to improved performance of built facilities depends on the behaviour and practices by clients and users. Insights into the differences and similarities of these strategies, competences, capabilities and practices may provide both researchers and practitioners with, amongst other things, a better understanding of the scope for action, and valuable inspiration on different approaches to managing processes in different stages of the life-cycle of buildings and organisations (Haugbølle and Boyd, 2016).

The second challenge is related to decision-making processes and accountability. Methods to involve users and stakeholders in decision-making processes during procurement, construction and operation of built facilities are paramount. These methods may include different types of collaborative arrangements such as partnering and public–private partnerships which support a construction industry becoming more oriented towards user demands. In turn, this requires working with greater transparency and accountability. In addition, managing differences between user needs and organisational objectives of clients may well deliver more value for money to both clients and users, but it is also likely to be challenged by asymmetrical information, conflicting interests, etc. (Haugbølle and Boyd, 2016).

The third challenge relates to learning and competence building. It is paramount to learn from successful/failed projects and investigate why some tools, procurement routes, etc., may be more appropriate than others under different circumstances in order to deliver value. Such studies are instrumental for managing information effectively, supporting organisational learning processes and developing appropriate guidance material for client and user organisations (Haugbølle and Boyd, 2016).

Part II of this book addresses issues related to how clients exercise governance, that is, the processes and mechanisms through which clients and users are shaping and being shaped by the supply system. This part consists of five contributions which cover:

- the funding of construction projects;
- defects and construction insurance;
- construction management capabilities of clients;
- client-learning across projects; and
- factors determining customer satisfaction and project quality.

The first contribution within this theme is by Abdul-Rasheed Amidu, who investigates the implications for clients of different financial arrangements for construction projects. Construction projects are by nature complex and dynamic human endeavours that involve the dilemma of funding them in an efficient and profitable manner. This chapter presents financing models for construction projects and explores their implications for clients. Although primarily focused on private projects, more and more public projects are procured using these models. Funding public projects through the Exchequer still requires a business case to be produced and a justification for the money delivered. However, the greater requirement for accountability of public funds restricts the freedom to act efficiently and increases the administrative burden throughout the project in order to deliver accountability. In particular the chapter considers the risk implications of funding public–private partnerships. The chapter does have a UK focus where substantial amount of public construction takes place under a private model. In many other countries, major construction is funded directly by the government where the delivery of the project is of greater importance than its financial efficiency. In these countries, however, as governments seek greater control, tighter financial approaches will be instigated.

The second contribution, by Kim Haugbølle, explores different approaches to defects and construction insurance and possible implications for clients: defects pose huge challenges to both clients and the construction industry. One important approach in dealing with risks associated with defects is guarantee policies after handover of a built facility – known as inherent defects insurance (IDI). This chapter will analyse the typical approach of inherent defects insurance and contrast this with an alternative approach as exemplified by the Danish Building Defects Fund. Based on a constructivist perspective, this chapter will identify two distinctive framings of construction insurance as either a protective mechanism or as a driver of change, and discuss the possible implications for clients.

The third contribution is by Youngsoo Jung and Seunghee Kang, who develop a methodology for assessing the construction management capabilities of clients. Although a client can strongly shape the primary directions of a construction project, literature on construction management has rarely investigated the characteristics of the clients' in-house construction management functions vis-à-vis outsourcing to construction service providers. Different types of client organisations may require different levels of involvement in the construction process. Improved understanding of a client's construction management capabilities, therefore, can facilitate more efficient planning of new construction projects for clients as well as construction service providers. Based on an exploratory study of Korean construction clients, this chapter presents a methodology for measuring the construction management capabilities of a client organisation based on fourteen construction business functions and for assessing the gap between current and required capabilities.

The fourth contribution within this theme, by Leentje Volker and Mieke Hoezen, deals with how major public infrastructure clients learn across consecutive infrastructure projects. By adapting the existing governance structures into new project arrangements, large construction clients learn from the preparatory and procurement stages of a project. This research is based on three large complex infrastructure projects of the Dutch Highways Agency. The results address the need for sense-making in the procurement stage of a project as an essential step in establishing a cooperative relationship between client and contractor. By learning from previous projects across and among sectors, clients can contribute to creating a sufficient governance structure for their future projects.

The fifth and final contribution within this theme is by Chimene Obunwo, Ezekiel Chinyio, Subashini Suresh and Solomon Adjei. Considering road construction projects in Nigeria, identification of the key sources of client satisfaction would provide a huge step towards redressing the chronic problem of client dissatisfaction. Based on a quantitative survey of 116 construction clients, Stepwise Multiple Regression Analysis was employed to evaluate the interrelationship between aspects of project quality (i.e. performance, reliability and aesthetics) and client satisfaction (measured through contractor referral and repatronage). The analysis revealed that project aesthetics was the most significant determinant of their satisfaction being alone responsible for 46 per cent of the variance in contractor repatronage and 65 per cent of the variance in contractor referral. The findings imply that, in order to satisfy clients, contractors need to enrich the aesthetic components of the constructed facilities. These findings have implications for other construction clients especially where public services are involved.

Part III: Innovation – change versus stability

The CIB W118 Research Roadmap (Haugbølle and Boyd, 2013, 2016) highlights a third overarching research and development theme to be addressed in relation to clients and users. This third theme is innovation, which is generally considered to be the key driver of improved wealth and quality of life. However, clients need to be able to manage a number of underlying challenges in relation to client and user practices and industry practices so that mutual benefits can be achieved. Hence with regard to change versus stability, the roadmap suggests a number of specific research objectives to be addressed (Haugbølle and Boyd, 2016: 6):

- exploring how clients and users can act as change agents of the construction industry;
- understanding the management of innovation as clients use buildings as instruments of change and new technologies impact on clients and users; and
- developing guidelines on how clients and users can support the move towards a sustainable future.

The first research objective here is related to risks and rewards for clients and users as change agents. In recent years, policy makers in many countries have advocated that construction clients take on a greater responsibility for stimulating innovation through (public) procurement of construction products and services. Stimulating innovation poses a dilemma for construction clients. While doing things in a different way potentially offers great rewards in time, cost and quality, it also inevitably increases risks. Hence, construction clients will often be caught between two different objectives: spending funds in a secure manner and/or taking on risks for adopting new technologies. This dilemma is particularly relevant for public clients with their heightened accountability with regard to spending public money securely (Haugbølle and Boyd, 2013, 2016).

The second research objective is related to the management of innovation, for example the nature of innovation activities, sources of innovation, the process of innovation and systems of innovation. As an intermediary between users and the construction industry, clients have to appropriately manage and balance the promises of new technologies against the risks. On the one hand, buildings themselves can be instruments of change when clients wish to change the operation of a client organisation. On the other, clients' setting their requirements influences which technologies, for example BIM, the construction industry adopts to deliver buildings and services. Either way, clients carefully need to manage the innovation process in an appropriate manner (Haugbølle and Boyd, 2013, 2016).

The third research objective is related to the course of development, more specifically the move towards a sustainable future. Clients play a particular important role with regard to procurement of sustainable buildings and refurbishment of buildings, while the behaviour of users has an equal importance with regard to the operation of buildings. Valuable contributions towards a sustainable built environment include providing insights, showcasing best practice and developing new guidelines and simulation for and with clients and users (Haugbølle and Boyd, 2013, 2016).

Part III addresses issues related to innovation in construction that deals with the relationship between change and stability. This part consists of four contributions that cover:

- a new integrated procurement methodology labelled three-envelope bidding system;
- technologies such as Building Information Modelling (BIM) as potential roads to business success for clients;
- seven different client roles in order to support innovation; and
- clients forming innovation networks for boosting change programmes in construction.

The first contribution is by Ada Fung and Ka-man Yeung and presents the development of a new procurement method developed by the Hong Kong Housing Authority. Construction developments commonly adopt the conventional design–bid–build procurement model where the design and construction functions are handled separately by different professionals with limited integration of expertise. This chapter examines an innovative three-envelope tendering system which strikes a balance between innovation and affordability, and maintains cost-effective use of resources in projects. This system capitalises on the potential of synergistic integration of stakeholders to participate collaboratively for achieving an integrated design and production process. It offers a leverage to drive the construction industry to research and innovate towards more efficient and sustainable construction.

The second contribution, by Niraj Thurairajah and David Boyd, considers Building Information Modelling (BIM) as a business opportunity to integrate building and its facility with the client's business processes. This is an explicit shift away from seeing BIM as being just a tool of building delivery, to one of being part of the business resource of organisations; this is key to this development. Construction industries around the world have started establishing national BIM programmes to address governments' needs to improve building delivery through the use of sharable information. In particular, public authorities have the power to enforce these changes, but private clients are also starting to see the benefits of using BIM and its potential to connect to a wider network of buildings and smart cities. Thus, clients are starting to realise the potential of integrating BIM with their business processes to build a competitive advantage via better value creation. This chapter explores clients' use of BIM and the challenges associated with its implementation. It proposes three digital dividends of BIM, accessibility, efficiency and transformation, and shows how these require the support of standards, skills and mutual development.

The third contribution, by Kristian Widén, expands our understanding of the role clients may play with regard to innovation and is based on a case study of a public client implementing building information modelling. Clients are important for innovation and innovation diffusion in the construction sector. Clients may be responsible for seven different roles or activities associated with promoting innovation:

- source/provider of knowledge;
- effective leadership;
- change agent;
- provision of financial incentives;
- appropriate forms of procurement;
- improved risk management; and
- disseminator of innovations.

This chapter features the analysis of a case in which the client took the lead in developing and diffusing BIM in the national construction sector, taking on all seven innovation roles and managing to influence the rest of the sector. The chapter shows that the activities associated with these roles really matter.

The fourth and final contribution within this theme is by Kim Haugbølle, Stefan Christoffer Gottlieb, Niels Haldor Bertelsen and Peter Vogelius. It is based on four previous studies of three different client innovation networks in Denmark and Sweden:

- the Swedish BeLok network for energy savings in commercial buildings;
- the Danish network AlmenNet for a broad range of development activities in the Danish social housing sector; and
- the Swedish BeBo network for energy savings in the residential sector, which was studied twice with a five years interval.

This chapter investigates how clients can organise themselves in innovation networks, establish collaboration with private and public funding bodies, and initiate a range of development activities. It assesses the impacts and benefits of clients forming such innovation networks and highlights some critical dilemmas and challenges that innovation networks may face.

Concluding remarks

It is our hope as editors that these contributions will stimulate further practice, research and theory development into the critical issues, questions and dilemmas raised in this book. In line with the CIB W118 research roadmap on clients and users in construction (Haugbølle and Boyd, 2013, 2016), particular emphasis should be placed on the three research themes of agency, governance and innovation.

We would like to thank the contributors to this book for their good spirit, colleagueship and patience in responding to the editors' repeated queries and comments. We hope that this joint effort by the contributors and editors will provide readers with new insights. As with all the volumes in Routledge's CIB Series, our goal is no less than to further our knowledge on and practices of current and pressing matters of great complexity.

References

Haugbølle, K. and Boyd, D. (2013). *Clients and Users in Construction: Research roadmap report*, CIB Publication 371. Rotterdam: CIB. Available at: http://site.cibworld.nl/dl/publications/pub_371.pdf (Accessed 28 August 2016).

Haugbølle, K. and Boyd, D. (2016). *Clients and Users in Construction: Research Roadmap Summary*, CIB Publication 408. Rotterdam: CIB. Available at: http://site.cibworld.nl/dl/publications/pub_408.pdf (Accessed 28 August 2016).

Part I

Agency

Roles and responsibilities

1 The merits of client associations

Henrik Lindved Bang, Marleen Hermans,
Rolf Simonsen and Karen Mogendorff

Introduction

Construction clients and client associations

This chapter focuses on the role and function of a relatively new phenomenon: client associations in the construction sector. The term 'construction client association' is used as a common denominator for construction client networks, associations and forums. There are not many client associations worldwide, though there are a number of property associations, but these rarely focus on construction.

Investigating the evolution of construction client associations and their agendas may provide insights and stimulate the professionalisation of client roles and client associations in the dynamic construction sector. The insights generated may be of interest to active clients in the sector that consider founding or participating in a client association. Thus, the aim here is to provide insights into how national client associations seek to contribute to the professionalisation of the construction client role in the dynamic construction environment. In doing so, this chapter intends to fill the knowledge gap identified in the CIB W118 Research Roadmap (Haugbølle and Boyd, 2013).

Construction clients are deemed to be important actors and potential change agents in the construction sector (Boyd and Chinyio, 2006; HM Government, 2013; Vennström, 2008; Vennström, 2009). The accomplishment, drive for change and innovative power of client organisations are of great consequence to the industry (Loosemore, 2002). The latter makes clients' professionalism an important element in leading and motivating change within the sector (Bonham, 2013). Individual client organisations professionalise to be better able to face challenges in the market. Additionally, construction clients increasingly organise themselves in associations to build their capacity to respond to major developments in the construction sector.

Current developments and challenges in construction and procurement

Construction clients associations aim to accomplish one or more of the following goals (CIB and CIDB, 2005; House of Commons, 2008):

- enable clients to get better value from the construction process in general and from procurement processes in particular;
- increase client competences to contribute to a better built environment through the procurement and commissioning of work;
- seek to influence government policy and the legislation process on behalf of their members; and
- increase the knowledge base for professional procurement and commissioning of construction works.

Commissioning in this chapter is defined as the way an organisation, in relation to its responsibilities in the built environment, shapes and implements its interaction with the supply market both externally and internally. Commissioning covers all activities relating to programming, selecting appropriate project delivery methods, setting the brief, procuring and contracting and contract management related to new construction, and managing the existing stock.

In order to reach these goals client associations may engage in the following activities or serve the following functions:

- provide networks for learning and exchange among clients;
- collect, promote and share best practices;
- collaborate on and participate in research and development projects and programmes;
- develop and provide a portfolio of products and services for members; and
- articulate a collective client voice in consultations with public authorities and the industry.

Client associations need to address a number of international trends in order to support the professionalisation of their members and the interests of the construction sector more broadly. Among these trends is an increasing internationalisation of the construction sector in Europe. National borders have been opened up, and the construction industry is becoming gradually more subject to European policies and regulations. A major development in the sector is the introduction of EU Public Procurement Directives (the latest in 2014), which directly influence public commissioning. The new directive introduces guidelines and regulations that promote a single open market, enhance innovation in public procurement and stimulate value-based procurement methods.

Additionally, the financial crisis hit the construction sector hard (EUROSTAT, 2013). The swift reduction of new construction work has forced many companies to shift their focus from new building to renovation of existing buildings and infrastructure and to overseas markets. Furthermore, with an aging, ill-performing building stock, sustainable transformation has become a key issue (European Commission, 2012). The challenge of sustainability induces greater attention on the interplay between buildings, cities and infrastructure leading to a shift in focus and activities of construction client organisations.

Furthermore, new project delivery methods, such as public–private partnerships (PPP), private finance initiatives (PFI), build–operate–transfer (BOT), design–build–finance–maintain (DBFM), new engineering contract (NEC) and best value procurement (BVP), develop rapidly within the construction sector partly due to the diffusion of new information and communication technologies. These developments incur new questions related to functional specification, competences required, organisational structures, financing, contract management issues and legal concerns. For instance, building information modelling (BIM) enables new project delivery methods and requires new competences.

Methodology

The comparison of client associations presented here is the result of co-operation between the Danish Association of Construction Clients (in Danish: Bygherreforeningen) and the Chair of Public Commissioning in Construction at Delft University of Technology in the Netherlands. The latter chair was established by the Dutch Construction Clients Forum (in Dutch: het Opdrachtgeversforum). Warren Parke in New Zealand, Tommy Lenberg in Sweden and Don Ward in the UK have also provided valuable inputs on the construction clients associations in their respective countries.

Description and comparison of how client associations promote professionalisation as presented in this chapter is based on a combination of desk research and consultations with representatives of the five client associations. The desk research included consultation of scientific literature on client associations, general literature on developments in the construction sector and available documentation on client associations such as written documents and web pages. Specifically, the desk research made it possible to place the founding and development of client associations in a wider context and helped to identify the relevant dimensions for comparison of the client associations.

Following the desk research, representatives of the five client associations were asked how much they agreed with the list of dimensions compiled for the comparison of client associations. Agreement was reached on a set of dimensions that are relevant and important for describing client associations.

In a following step the same representatives were asked to provide information about each client association on each of these dimensions. For each of the five associations, data was collected on the following dimensions:

- short history of origin;
- main aims and objectives;
- structure, membership base, ways of working and organisation;
- funding and financial structure;
- main activities, products and results;
- themes and strategic issues addressed;
- affiliations to government, trade organisations, R&D programmes and universities;

- major developments since establishment;
- future plans and ambitions; and
- an inspirational story that highlights challenges and successes of the association.

Based on the feedback of the client associations, tables were compiled to enable a systematic comparison of associations on the dimensions identified in the desk research. The tables were then sent to the representatives for verification.

Typology and professionalism of clients

Client associations represent the interests of construction clients: a very diverse group of organisations with substantial differences in the conditions they operate in and their characteristics. In this section, typologies of construction clients are described to the extent they have a bearing on the professionalisation of client roles through associations and inform the analysis of client associations presented later in the result section of this chapter.

Typologies of construction clients

According to the Organisation for Economic Co-operation and Development and Eurostat (OECD and Eurostat, 1997: 11) a construction client is '*the natural or legal person, for whom a structure is constructed, or alternatively the person or organization that took the initiative for the construction*'. Several authors have tried to classify types of client (Chinyio *et al.*, 1998). Elements covered in existing typologies are: experience (experienced versus inexperienced), legal status (e.g. public versus private) and diversity in building projects, including whether or not a client constructs and manages assets for its own use. Ideally, these typological elements affect the required and expected professionalism of the client involved.

One of these typologies is related to experienced versus inexperienced clients. A specific challenge to professionalisation of the construction industry is that most organisations act only occasionally or even just once as a construction client (Chinyio *et al.*, 1998). Only a minority of construction clients can be considered professional in the sense of employing full-time staff specifically for procurement of construction and asset management related activities. Johansson and Svedinger (1997) provide a workable distinction between three types of construction client – user-client, manager-client and seller-client – with reference to an organisation's experience with procuring construction work and asset management services. The user-client is an organisation owning and using its own premises and acting as a construction client for these premises only. A manager-client is the 'asset management' type of client, taking care of a portfolio of built stock and thus being a repetitive construction client. The seller-client develops stock for the purpose of selling it after completion. Commissioning construction work is a core task of the organisation only for manager-clients and seller-clients. Put differently, the frequency of the activity is related to the type of client. Examples of professional client organisations are housing associations, government building agencies, infrastructure authorities and municipalities/regions.

The second typology is related to the legal status of clients. An important distinction made by Boyd and Chinyio (2006) is the one between public, private or public–private client organisations. Public organisations differ from private clients in that they are compelled to respond to specific procurement regulations and to adhere to public values (De Graaf and Paanakker, 2014). Both procurement regulations and public values tend to be subject to politically driven change and affect or are affected by societal debates, for example about privatisation (de Bruijn and Dicke, 2006; Reynaers and De Graaf, 2014). Given that the knowledge, competences and domain of operation may differ between private, public, non-governmental or not-for-profit clients, the activities of client associations may be partly explained by whether they represent all construction client organisations or a specific category.

The third typology is related to the type and diversity of construction activities commissioned by clients. The construction sector is a heterogeneous field. Different types of construction activity tend to have their own distinct playing field and require specialised knowledge and specific competences, for example it matters whether clients engage in civil engineering, house building, refurbishment or some other activity. Another distinction is whether the client is building for its own use or for another organisation – for example the client is a developer who constructs buildings for resale. There are intermediate forms where the client organisation is more or less detached from the facilities management organisation although they belong to the same (parent) organisation.

Aspects determining the professionalism of construction clients

We describe the rich diversity of the potential membership base of client associations in general. The types of clients represented by client associations may affect the goals and activities of client associations. The membership base is therefore a key dimension on which the five participating client associations are compared in the result section. In the following subsection we discuss the different aspects of the day-to-day work construction clients engage in and that client associations seek to professionalise: aspects of professionalism with regards to the procuring activities related to buildings and infrastructure.

Eisma and Volker (2014) concluded from a systematic literature survey on aspects related to (public) commissioning that the concept of 'commissioning' with regard to professionalisation refers to both the construction client's role in projects or maintenance assignments and to organisational issues. Further research by Hermans *et al.* (2014) helped to identify ten core elements that are important for organisations seeking to professionalise or transform their role as a construction client. Each element reflects a specific set of capabilities construction clients may need to conduct theirs tasks professionally. Two of the identified elements are particularly relevant to public clients: the ability to apply public values in procuring activities, and the ability to adhere to public rules of play, such as procurement regulations. The eight other aspects apply equally to public and private construction clients.

According to Hermans *et al.* (2014), the 10 aspects defining professionalism in the role of public construction clients are:

- organisation strategy and policy;
- culture and leadership;
- people and competence management;
- deliberate use of decision models and portfolio management;
- relationship with stakeholders;
- conceptualisation of public role;
- engagement with public rules of play;
- interaction with supply chain and market;
- managing projects and assignments; and
- creativity and flexibility in applying aforementioned aspects.

Based on these 10 aspects an inventory was made of the services client associations offer to enhance construction clients' competences. The ten aspects and the results of this inventory are used for further analysis in the following sections.

Characteristics and roles of five national client associations

In this section an analysis of the similarities and differences between national construction client associations is presented with regard to how they seek to professionalise client roles. The section starts with the founding of and the main characteristics, aims and activities of the five associations investigated.

Establishment of associations

The establishment of each of the five client associations was based on a shared vision of construction clients with regard to their professional responsibility, the need for reform of the sector and the necessity of co-operation aimed at sharing knowledge and experience in order to influence clients' framework conditions. Awareness has grown that professionalisation of the construction sector requires that specific needs and conditions are met. Construction clients cannot all be expected to meet these needs and conditions on their own. This prompted the founding of construction client associations in various countries: Bygherreforeningen in Denmark, Opdrachtgeversforum in the Netherlands, Construction Clients Group (part of Constructing Excellence) in New Zealand and Construction Clients Group (part of Constructing Excellence) in the UK. In the UK, the Construction Clients Forum was set up as an independent body in 1995. Since 2005 the Construction Clients Group has been an independent body under the Construction Excellence programme. Moreover, most associations were established around the turn of the millennium – only the Swedish Byggherrarna is older (see Table 1.1).

Table 1.1 Characteristics of client associations

	Bygherreforeningen (DK)	Byggherrarna (SE)	Opdrachtgeversforum (NL)	Construction Clients Group (UK)	Construction Clients Group (NZ)
Year of establishment	1999	1964	2004	1995/2005	2004
No of members	120	133	12	30	40
No of board members	9	13	12	3	?
Membership base					
Public client members	X	X	X	X	X
Private client members	X	X		X	X
Domains					
Civil engineering	X	X	X	X	X
Housing	X	X	X	X	X
Non-residential	X	X	X	X	X
Types of clients					
User-clients	X	X	X	X	X
Manager-clients	X	X	X	X	X
Seller-clients	X	X		X	X

Membership base

Generally, the membership base consists of the large public construction clients who want to strengthen their collective influence. But every rule has its exception: in Sweden a group of large private industrial clients facing a strong contractor industry founded the association. However, over time both public and private clients tend to join the association. Manager-clients tend to be the first to join associations, while user-clients and seller-clients tend to join these associations later and in smaller numbers. Moreover, most segments of the construction industry tend to be covered by the membership base, for example infrastructure and utilities, commercial and administrative buildings, and housing and residential.

Notwithstanding the similarities between client associations, the structure and member focus of the individual associations varies considerably. The Dutch association has only (semi) public clients as members whereas other associations also have private clients. The UK Construction Excellence programme was from the beginning largely funded by the government, whereas the Scandinavian and Dutch associations were primarily financed by members. The Dutch forum can be characterised as a board member network focused on professionalisation of the sector, whereas the Scandinavian associations focus on knowledge development and training of all commissioning related functions within client organisations.

Legal structure and funding

Data on the individual client associations indicate that the legal status of the examined associations appears to matter. Only associations that are also associations in legal terms, appear to be affiliated with other trade associations and/or client organisations (marked as 'Other' in Table 1.2). The latter suggests that legal structure may affect which organisations associations choose to work and affiliate with.

Some of the client groups partly operate as trade organisations and through this role seek to influence legislation and engage in dialogue with government and industry. Others limit their role to knowledge dissemination.

There is also a notable difference in the funding of associations. Membership fees tend to be an important source of income, although the funding may have been different at the establishment of the associations. However, the client associations in the UK and New Zealand are sharing facilities with larger organisations and are thus benefiting (through synergies) from funding from this

Table 1.2 Legal structure and affiliations

	Bygherreforeningen (DK)	Byggherrarna (SE)	Opdrachtgeversforum (NL)	Construction Clients Group (UK)	Construction Clients Group (NZ)
Legal structure					
Association	X	X			
Network			X		
Other (part of CE)				X	X
Funding structure					
Membership fees	X	X	X	X	X
Contracts with agencies		X			
Foundations and public dev. programmes	X				
Educational programmes/activities	X	X			
Other				X	X
Affiliations					
Government	X	X	X	X	X
Universities		X	X		
Others	X	X		X	X

larger organisation (Constructing Excellence). Both Construction Clients Groups are governed by separate boards to preserve their independence in policy and activities as client associations. There is a delicate balance concerning the perceived independence of client associations with this type of affiliation and funding.

Organisational structure

A client association generally has a steering committee (board) consisting of top managers – who reflect the membership base – and a secretariat taking care of the association's day-by-day management tasks. Associations also tend to evolve over time from informal networks to established associations or trade organisations, including working groups organised around specific subjects.

Moreover, all client associations are affiliated to government agencies and universities in one way or another, irrespective of their legal status. There are also notable differences. Affiliations with university are especially strong in the Dutch Construction Clients Forum (Opdrachtgeversforum). In 2014 this network established a chair in public commissioning – a scientific research group headed by a professor – at Delft University. The group also hosts the secretariat of the Dutch Construction Clients Forum. In contrast, the Danish client association is hosting the secretariat of a general reform programme (Værdibyg), and, in the UK and New Zealand, the client association is being hosted by a general reform programme, namely Constructing Excellence.

Activities

All associations primarily seek to empower their members, by focusing on increasing relevant competences, producing services and products for members, and providing opportunities for networking. Most of the client associations collaborate with universities or research institutes on R&D projects and programmes. These efforts primarily seek to create better value for money and to improve procurement practices.

Another important function of client associations is to give the construction client community a collective voice in their dealings with both public authorities and other construction industry associations. For some public clients it is possible to communicate their viewpoint more freely and forcefully through a client association than through their own contacts. All associations seek to influence government policies but they go about it in different ways and to a different extent. For instance, the Swedish Construction Clients Association has a legal advice service for its members. It may be that the Swedes, as the oldest association by far, are forerunners in this respect. It is also of interest that some associations – in particular the Dutch Construction Client Network and the UK Construction Clients Group – seek to improve the reputation of the construction sector as a whole. Table 1.3 provides an overview of the main activities of the five construction client associations.

Table 1.3 Activities of construction client associations

	Bygherrefreningen (DK)	Byggherrarna (SE)	Opdrachtgeversforum (NL)	Construction Clients Group (UK)	Construction Clients Group (NZ)
Increasing competences	X	X	X	X	X
Better value for money/better procurement	X		X	X	X
Providing a network among clients	X	X	X	X	X
Collect and share best practices	X			X	X
Collaborate on R&D projects and programmes	X	X	X		
Develop and provide products and services	X	X	X	X	X
Influence government policy and legislation	X	X		X	X
Emanate collective client voice	X	X	X	X	X
Change values/mind set/processes in procurement	X			X	X
Offer legal advice to members		X			
Enhance the reputation of the industry				X	X

As illustrated by Table 1.3 the five forerunners serve largely the same functions. It seems plausible that the client associations continue to develop new services when they mature and in response to emerging challenges; examples of such services may be the offering of legal services and the explicit aim to improve the reputation of the construction sector as a whole.

Engagement with sectoral challenges

With regard to engagement with sectoral challenges, the five associations tend to focus their activities on procurement regulations and new types of collaboration, project delivery methods and standard forms of contracts. The latter may be seen as a response to major economic, political and societal developments and changes that affect the construction sector both in and outside of Europe. Sustainability and to a lesser extent health and safety and labour flexibility are important themes that have increasingly become of interest to client associations. For some associations these issues are already well-established themes or part of the package of services and products offered by the associations. Typically, associations establish task forces to work on specific themes. An overview of themes covered by the client associations can be found in Table 1.4.

Table 1.4 Client associations' priority themes

	Bygherreforeningen (DK)	Byggherrarna (SE)	Opdrachtgeversforum (NL)	Construction Clients Group (UK)	Construction Clients Group (NZ)
Internationalisation	X				
Procurement regulations	X	X	X	X	X
New types of collaboration and contracts	X	X	X	X	X
Information technology and BIM	X	X	X	X	X
Labour flexibility		X			
Increased emphasis on existing assets	X		X		X
Sustainability	X	X	X	X	

Service provision for professionalisation of clients

Associations tend to address many of the aspects deemed important for construction clients' professionalism. The aspect 'public values' is, however, not mentioned in most work programmes of the associations investigated. A likely explanation is that four out of these five associations – the Dutch association is not included here – do not focus exclusively on public construction clients. 'Creativity and flexibility' are not exclusive to the domains of public construction clients, or even construction clients as a whole, but are probably relevant characteristics for all entities dealing with complex surroundings and multifaceted working activities.

The specific focuses of the client associations investigated differ in some respects, their differences associated with the structure of the industry, institutional setup and sector development in each country. Nearly all associations seem to have a broad scope of activities and themes with the exception of the New Zealand association, which appears to be more focused on project-related issues. Table 1.5 compares the services provided by client associations with the ten aspects of professionalism addressed earlier.

Table 1.5 Comparison of aspects of commissioning and client associations' service provision

	Bygherreforeningen (DK)	Byggherrarna (SE)	Opdrachtgeversforum (NL)	Construction Clients Group (UK)	Construction Clients Group (NZ)
Organisational strategy and policy	X	X	X	X	X
Culture and leadership		X	X	X	X
People and learning organisations	X	X	X	X	X
Decision models and portfolio	X		X	X	
Stakeholders	X	X			
Public values				X	
Public rules of play	X	X	X		
Interaction with the supply market		X	X	X	X
Managing projects and assignments	X		X		X
Creativity and flexibility		X		X	

National client association development stories

Every construction client association operates in a distinctive national context. In this section, factors are described that representatives of client associations consider central to the functioning of their association. These national stories may be useful to construction clients that seek to start their own association.

Denmark: hosting development initiatives and extensive industry collaboration

The Danish Association of Construction Clients (DACC) is a key player in the development of the national construction sector. DACC hosts and leads development programmes and projects. These activities are funded by private foundations and public development programmes and account for more than half of the association's activities and budget. One of the areas that DACC currently explores is value creation and user involvement in construction projects and the production of guidelines and web tools for better briefing of projects.

Key to achieving influence and affecting change in the Danish construction sector is DACC's close collaboration with other professional associations responsible for the interests of contractors, engineers, architects, suppliers and construction workers. One recent example of DACC's success is the introduction of a common charter of Corporate Social Responsibility (CSR) across the construction industry. The charter led to the establishment of a new association for CSR in construction which is hosted by DACC. This has been achieved in cooperation with other professional associations and shows how the industry itself can make a difference with regards to responsibility and reputation.

The Danish professional construction associations have under the auspices of DACC founded a development programme Værdibyg (the Value-creating Construction Process) focused on improving construction processes for the benefit of all actors in the sector. A prime success is the acknowledgement that these various actors need to work together to achieve results. The programme produces shared guidelines for construction processes on, for example, project collaboration, project optimisation, procurement strategies, commissioning and scheduling. Additionally, the secretariat for Lean Construction Denmark is hosted by DACC.

The Netherlands: executive network and university chair

The Dutch Construction Clients Forum (DCCF) focuses on construction clients from the public domain: government agencies, the railway agency, municipalities, hospitals, housing associations, water agencies and the like. DCCF is a non-institutionalised, informal network organisation with a small steering group aiming specifically at interpersonal knowledge exchange on board level. Each board member represents a particular segment of the sector. The personal commitment of the board members towards DCCF activities is considered to be one of the most important assets of the Forum; it ensures a 'hands-on' approach towards DCCF initiatives and stimulates exchange of good practice between segments.

DCCF has over the years developed a clear voice and has become a trusted contact for issues related to commissioning. Both industry and policy makers have approached DCCF whenever they need to discuss developments and align (public) construction clients' opinions on and support with regard to these developments. An example of the latter is high level summits of board members of construction clients and large Dutch contractors.

A number of successful initiatives were launched or supported by DCCF, such as a joint Project Academy that seeks to train senior project managers involved in complex construction projects. Other accomplishments include, but are not limited to, a code of conduct on professional commissioning, which has been adopted and applied by most segments represented by DCCF, and a joint tool for implementing social return in procurement processes. A promising new initiative is a tool that measures past performance of the building industry as a basis for future tendering processes. Through large-scale events, DCCF involves the various segments represented by steering group members. The annual 'Dinner Pensant' serves as a basis for discussing the state of the art in public commissioning and exchanging visions on aspects related to public commissioning. Through these larger-scale events, and sector specific activities organised by the individual steering group members, the DCCF is firmly 'grounded'. Due to the fact that DCCF is a small informal organisation, the impact of DCCF is mainly to advocate and promote the commissioning role.

The establishment of a chair on public commissioning at Delft University can be seen as a major accomplishment for DCCF and is unique in European academia, though it is more common with industry sponsored chairs in the US. Through its research and education programme, the chair enforces the visibility of construction clients and their activities, fundamentally adds to the knowledge base on professionalisation of commissioning, and supports the development of educational programmes.

United Kingdom: benchmarking and client commitments

Professional UK clients have been pioneering the use of key performance indicators (KPI), which have enabled organisations to benchmark their performance against the rest of the construction industry on issues that are critical to the success of projects and organisations. Increasingly, Constructing Excellence's construction industry KPIs are being utilised by both public bodies and the private sector in the drive towards continuous improvement.

The UK Construction Clients Group (UK CCG) has been successful in getting clients to consider their business needs before commissioning construction and, having initiated a project, to have high and appropriate expectations of the entire project team, including themselves. This has been driven by the formulation of so-called Client Commitments to performance improvement. Participants measure their progress against an agreed programme with increasingly demanding targets, and the companies that meet the targets are rewarded with continued business. The Client Commitments focus on six areas where clients can make a positive difference to enable better value:

- client leadership;
- procurement and integration;
- health and safety;
- design quality;
- sustainability; and
- commitment to people.

These principles were demonstrated successfully on a large scale during the construction of the facilities for the 2012 Olympic Games in London.

UK CCG is active in influencing government policy and legislation through high-level consultations with industry and government on a number of task groups and in collaboration with a host of organisations, such as Strategic Forum for Construction, Construction Skills, CIOB, HSE and other stakeholder groups.

New Zealand: National Construction Project Key Performance Indicators

The New Zealand Construction Clients Group (NZ CCG) predominantly mirrors the UK model and has achieved 10 years of operation in New Zealand. The major achievement of the NZ CCG has been the adoption and refinement of the National Construction Project Key Performance Indicators. Members now have access to suites of performance data for 2006, 2011, 2012 and 2013. This has enabled members to measure their project performance and identify trends or areas to address as well as compare themselves with not only the NZ industry but with the UK as well – a powerful business improvement tool.

It is a national attribute for New Zealand to compare and learn from what is happening in the UK. Accordingly, through this awareness, New Zealand can often avoid the 'pitfalls' of new ideas and technologies already in place in the UK and internationally. By necessity, New Zealanders are innovators as well, and export innovations commercially. With the information age speeding up the dissemination of ideas globally, the lag with the UK and the rest of the world is narrowing.

The depth of the data in the National Construction Project KPIs has been of great interest to the government and universities as it has been operating throughout the global financial crisis and can show trends and allow predictions to be made on how the industry is expected to perform under the 'boom–bust' cycle that affects both the NZ and the UK markets.

The strong working relationship NZ CCG has established with government officials helping regulate the industry allows NZ CCG to be at the forefront of changes and developments that the government pursues. The impacts of the 'boom–bust' cycle, the leaky building legacy of the 1990s and the Christchurch City rebuild since the 2010 earthquakes have created pressures on the NZ construction industry. Certain 'conservative' behaviours regarding risk apportionment in the market has created negative contracting and wasteful outcomes. The NZ CCG has been afforded the opportunity to offer advice and influence how the government is addressing this issue, and this demonstrates the benefits of solidarity and 'lined-up' thinking in shaping decision makers.

Sweden: successful lobbying and client masterclass

One of the most important success factors of the Swedish Construction Clients (SCC) is its extensive network and its ability to utilise the network to keep developing efficient tools, collaborations and education. These assets facilitate lobbying in a way that is beneficial for the members of SCC. It is through cooperation with several other key operators that the SCC productively contributed to the development of a best practice tool for measuring the results of a construction project.

Swedish Construction Clients has been committed for some years to the development of the construction sector. This has been done by setting up and taking part in a number of programmes for research and development aimed at strengthening the industry. Some of the results emanating from projects and studies within these R&D programmes have been published internationally. Moreover, Swedish Construction Clients is a highly valued partner with the Swedish Energy Agency and has for several years been responsible for three of the agency's networks on energy savings.

As in Denmark, SCC has also been successfully focusing on their construction clients' education and development of skills. During the past 10–15 years a number of training and education courses have been developed in cooperation with partner organisations. 'Construction Client Masterclass Diploma' is an MBA certificate that has been developed at the Swedish technical universities in close cooperation with SCC, who had the hands-on knowledge of the needs of the construction clients. Today, Construction Client Masterclass is an executive programme offered by IFL Executive Education at the Stockholm School of Economics.

Success factors: lessons to be learned

From the comparison of client associations, conclusions may be drawn and lessons learned regarding how these as organisations contribute positively both to their client members and the construction sector and society in a broader sense.

A healthy organisational basis

When setting up a client association, a healthy basis in terms of the initial core membership base is paramount. The founding members need to be large and influential and with representation from senior management or board level, where each representative should be personally committed to the association. Their views on the need for change must be closely aligned, which takes plenty of effort and work in itself.

Once this healthy basis is established the membership base may be expanded in many ways as developments pick up pace. Hence, both Swedish and Dutch cases are examples of very determined and focused beginnings (private versus public focus) that have laid solid organisational foundations for activities. Yet, even in the UK and New Zealand, which from the beginning had other types of organisations closely linked to the associations, there is a clear understanding that the core of client membership needs to be strong and focused. A strong grouping of clients will attract other potential members and interest groups and decision makers.

A good working relationship with government

From the outset a new client association needs the acceptance of government to gain momentum and secure credibility. For most client associations it is important and natural to present organisational objectives that address societal concerns, such as regarding improved quality, productivity and value creation of the construction sector. This also explains how several client associations have been established as part of or following governmental reform programmes addressing the role of demand-pull in improving the performance of the sector. Construction client associations need to nurture a good working relationship with relevant government departments that govern the industry.

Client members want a voice and a channel to approach government and other decision makers. Client associations can present concerns and comment on developing legislation when public comment is sought. The same applies when government agencies want to engage with representative groups of the construction industry. Alliances between client associations and other industry organisations are important to gain influence. Some of these activities take place through informal consultations, while others are formalised in committees, hearings, etc.

Relevant network and targeted communication activities

Most client associations are active in both network and communication activities aimed at the membership base. In some cases this is restricted to high-level management (the Netherlands) whereas in others the spread is very large in certain types of participant (open seminars and specialised networks in Denmark) and regional activities (Sweden and the UK).

The role of the secretariat of client associations is very different in seeing that these activities are relevant and timely. Websites and newsletters are used in most cases as well as social media such as LinkedIn and Twitter. For all these activities, it is important to pick topics that are highly relevant to members and select each topic to be guided by leaders in those fields, for example innovative and value-adding approaches. It is often a challenge to target the interests of different kinds of client organisation and the level of individual members (e. g. project managers versus executive managers).

Level of competence and development of tools

Every client association is in some way or another engaged in improving the level of competence of clients to raise professional standards. Some are heavily engaged in university research and develop courses for managers (the Netherlands and Sweden), whereas others have widespread programmes of various short courses and project management modules (Denmark). Others have established 'Client commitments' and benchmarking systems that set standards for 'good client-ship' (Sweden and the UK).

Most client associations have active work groups in areas of particular interest that contribute to both policy formulation and development of new tools and methodologies for clients. Some client associations have more than 100 persons engaged in work groups and more than half of their annual budget from externally funded projects developing new tools and methodologies (Denmark).

Perspectives on challenges in the future

Summing up, the characteristics and activities of the five construction client associations has hopefully demonstrated that they play an important role, which may in turn act as an inspiration for others who wish to develop a strong client base on a national scale or internationally. Almost by definition, the future of national client associations is hard to predict, yet a few perspectives are offered here.

The areas of focus for client associations have to a large degree confirmed the theory. Client associations are concerned with professionalisation, increasing levels of knowledge and expertise, and gaining influence for clients. However, the client organisations are quite different with regard to their needs and affiliations. It will be interesting to see if these differences put pressure on the integrity of client associations as they grow in breadth in both membership base and activities offered to members. It will also be interesting to see how the interaction with bordering associations such as those of the property profession and facilities management will develop.

Client associations are usually initiated on the basis of a tight-knit group of high-level people from some of the large member organisations. Most of the five client associations are still characterised by a strong involvement from these high-level 'grass roots'. This compensates for the fact that client associations are usually small in terms of members and with limited budgets compared to other industry associations in construction. It is an interesting question to ask whether this formula – where personal involvement and purchasing power of clients form a potent combination – can keep up with the growing political power and strong international network of other types of association. It is an open question if the strong personal involvement of founders will go on to future generations of key people in the client associations, or whether the client associations will be able to hold their power base through alliances with other organisations.

In times of increasing internationalisation it is remarkable how client associations – almost by nature – focus on the national level in terms of political decision making, legal and legislative frameworks, etc. Yet, increasingly many such issues are governed by international frameworks such as EU directives. Also, in terms of increasing value creation and innovative practices, a stronger international commitment seems necessary for clients. More research and development in client issues would also call for increased internationalisation. Accordingly, there seems to be a strong incentive for increasing the international perspective and engage in closer collaboration across borders (the strong links between New Zealand and the UK is an interesting example of this). Stronger international collaboration would not only benefit existing client associations, but may also be an impetus for establishing new client associations in other countries in the future.

References

Bonham, M. B. (2013). Leading by example: new professionalism and the government client. *Building Research & Information*, Vol. 41, (1), 77–94.

Boyd, D. and Chinyio, E. (2006). *Understanding the construction client*. Oxford: Blackwell.

Chinyio, E. A., Olomolaiye, P. O., Kometa, S. T. and Harris F. C. (1998). A needs-based methodology for classifying construction clients and selecting contractors. *Construction Management and Economics*, Vol. 16 (1), 91–8.

CIB and CIDB (2005). *Client best practice: an international perspective*. Discussion Paper. ICCF – International Construction Clients Forum, Sharing International Best Practice in Construction, 17–18 October 2005. Available at: www.jcp.co.za/articles_files/ICCF%20Discussion%20Paper.pdf (Accessed 27 Aug 2016).

De Bruijn, H. and Dicke, W. (2006). Strategies for safeguarding public values in liberalized utility sectors. *Public Administration*, Vol. 84 (3), 717–35.

De Graaf, G. and Paanakker, H. (2014). Good governance: performance values and procedural values in conflict. *The American Review of Public Administration*, Vol. 38 (2), 120–28.

Eisma, P. and Volker, L. (2014). Mapping fields of interest: a systematic literature review on public clients in construction. In: Jensen, P. A. (Ed.) (2014). *Proceedings of CIB facilities management conference: using facilities in an open world creating value for all stakeholders*. Lyngby, Denmark: Polyteknisk Boghandel og Forlag, pp. 129–43.

European Commission (2012). *Strategy for the sustainable competitiveness of the construction sector and its enterprises, Brussels, COM (2012)*. 433 final. Brussels: European Commission.

EUROSTAT (2013). *Construction statistics*, NACE rev.2. Available at: http://ec.europa.eu/eurostat/statistics-explained/index.php/Construction_statistics – NACE_Rev._2 (Accessed 27 August 2016).

Haugbølle, K. and Boyd, D. (2013). *Clients and Users in Construction. Research Roadmap Report*. CIB Publication 371. Rotterdam: CIB.

Hermans, M., Volker, L. and Eisma, P. (2014). A public commissioning maturity model for construction clients. In: Raiden, A. and Aboagye-Nimo, E. (Eds.) (2014). *Proceedings 30th Annual ARCOM Conference, 1–3 September 2014, Portsmouth, UK*, Association of Researchers in Construction Management Conference, ARCOM 2014, 1305–14. Available at: www.arcom.ac.uk/-docs/proceedings/ar2014-1305-1314_Hermans_Volker_Eisma.pdf (Accessed 27 August 2016).

HM Government (2013). *Construction 2025, Industrial Strategy: government and industry in partnership. URN BIS/13/955*. London: HM Government. Available at: https://www.gov.uk/government/uploads/system/uploads/attachment_data/file/210099/bis-13-955-construction-2025-industrial-strategy.pdf (Accessed 27 Aug 2016).

House of Commons (2008). *Construction Matters*, Ninth report of session 2007–08, V2, HC127-II, Ev. 76–82. Business and Enterprise Committee. London: The Stationery Office. Available at: www.publications.parliament.uk/pa/cm200708/cmselect/cmberr/127/127i.pdf (Accessed 27 Aug 2016).

Johansson, B. and Svedinger, B. (1997). *Kompetensutveckling inom samhällsbyggnad, Byggherren i fokus (in Swedish: Skills development in civil construction. Construction client in focus)*. Stockholm: Kungliga Ingenjörsvetenskapsakademien.

Loosemore, M. (2002). Impediments to reform in the Australian building and construction industry. *Australian Journal of Construction Economics and Building*, Vol. 3 (2), 1–8.

OECD and EUROSTAT (1997). *Sources and Methods: Construction Price Indices*. Paris: OECD. Available at: www.oecd.org/industry/business-stats/2372435.pdf (Accessed 27 Aug 2016).

Reynaers, A. and de Graaf, G. (2014). Public Values in Public–Private Partnerships. *International Journal of Public Administration*, Vol. 37 (2), 120–28.

Vennström, A. (2008). *The construction client as a change agent: contextual support and obstacles*. Doctoral thesis; No. 2008:31. Luleå: Luleå Tekniska Universitet.

Vennström, A. (2009). Clients as initiators of change. In: Atkin, B. and Borgbrant, B. (Eds.) (2009). *Performance Improvement in Construction Management*. London: Routledge/Spon Research, pp. 14–24.

2 A model of clients and users
A corporate real estate view

Christopher Heywood and Russell Kenley

Introduction

Any book on clients and users risks entanglement in the definitions of each and the knotted inter-relationships between them. As is the nature of many soft systems, the boundaries are vague and any clarity can appear to come from semantics rather than demonstrable facts. External models, arising from different perspectives, can provide a lens to help untangle the world-of-building clients and users. One such model, drawn from the literature of corporate real estate (CRE) using a demand and supply perspective, taps into an existing understanding of what it is to achieve value from constructed assets. We adopt CRE as the general term for the corporate real estate field as is often the case in the field itself, but note that there are important conceptual differences between CRE as real estate objects – physical properties and leases – and the practices required to manage them – corporate real estate management (CREM). The view offered by this model changes the conception of a client into a demand-side, real estate-using entity that has real estate and buildings to support their organisational purpose.

The CRE perspective introduces the 'CRE organisation' as a symbolic representation of 'clients' and 'users' – the CRE organisation is a client and has users. In this view the perspective shifts towards the organisation's business objectives and, to a lesser extent, its business relationship with users. This is based on the definition of CRE as:

> The real properties that house the productive or business activities of an organisation that owns or leases and, consequently, manages real estate incidental to its business objectives where the primary business is not real estate (after Rondeau (1992); Brown et al. (1993); Kenley et al. (2000); CoreNet Global (2007)).

The 'CRE organisation' is then one that has such property. Here it is not the investment value but the use value that is most important and, therefore, how the real estate supports core business objectives. This is not dissimilar to Boyd and Chinyio's (2006) observation that building (as an activity) is about organisational development. We agree, but expand the object of interest to be real estate and its entire life cycle of organisational use.

A CRE perspective is also useful because many of the 'problems' and issues related to clients identified in the W118 Research Roadmap Report (Haugbølle and Boyd, 2013) are things that the CRE domain addresses either theoretically or in practice. These include:

- the problem of clients not seeing themselves as clients but rather as being in business or offering services having a 'core business' typically outside of construction. In CRE this is equivalent to organisations seeing themselves as 'not in real estate' despite occupying many square metres of space in multiple locations and spending millions in occupancy costs. CRE's response has been to recognise the centrality of business purpose to the organisation and to argue that CRE is intrinsic to that business;
- the desire to make the client's purpose central to research into them makes CRE theory useful, as CRE alignment theory has a central concern of 'what is the organisation's purpose, and what maximises real estate's fit with that purpose?'. This makes strategic management of the organisation – where organisational purpose is fundamental – a central concern to CRE, likewise real estate's role in achieving that purpose; and
- the observation that clients are not unitary entities due to their differentiated organisational structures. A client could be a standalone business unit, senior management or individuals acting on behalf of the organisation. CRE offers several advantages here. First, the CRE function is itself an essential and intrinsic organisational function that deals with these differentiated structures every day. Those structures may constitute business units or other business functions such as IT, HR or marketing. To counteract this, when there is a centralised CRE function for the whole organisation – a key recommendation in CRE theory – it is possible to develop pan-organisational discipline in managing and providing real estate. This can result in a more unitary expression as a client.

This chapter expands the CRE theorisation introduced here to support the usefulness of CRE theory to the client and user problem and to establish the basis for introducing several CRE models. A case example is introduced so that the individual models can be illustrated before synthesis into a single model of the demand-side in relation to procurement activities.

Research methodology

CRE and the 'client and user problem'

A CRE theorisation may prove useful to understand the 'client and user problem'. It may not be a complete answer as CRE is just part of the real estate-using world, but it is a significant part because we argue that the only substantial alternative domain to CRE is private residential real estate. Because the CRE organisation lens provides novel thinking about clients and users, some more relevant background theory is useful to understanding 'client-ness' and for locating the CRE models that follow (for additional information on CRE, see Haynes and Nunnington, 2010; Edwards and Ellison, 2003).

Most organisations have real estate as a support to their business missions, when they act as construction clients and contain or serve many users (be they parts of the business, individual employees or customers). Though client and user relationships are complex, the CRE perspective suggests that their differentiation emerges from the organisation's business purpose. For example, an organisation which provides housing as a business purpose will include residents as users. A manufacturing business would include production operations as users. A bank is more complex but would include users that work in offices and in retail branch outlets providing financial services and also its customers, members of the general public. A bank may also have sub-organisational business units that are users but can also operate as clients.

In this chapter we focus primarily on clients while recognising that users are important in understanding the business purpose of real estate. We call this real estate using entity a 'CRE organisation' and distinguish that from the part of the organisation responsible for managing CRE which is its 'CRE function'. Under this conception there does not need to be an identifiable 'CRE Unit' because decisions about real estate use must be made by someone acting as the CRE function. This distinction is informed by functional and resource-based views of the firm, that is, organisations consist of various functions, such as operations, marketing, human resources and the like (Thompson and Strickland, 2003). These functions are deployed as standalone entities or configured into various business units. In the resource-based view, organisations control and use various strategic resources – land, labour and capital in classical economic terms, or, more contemporarily, human resources, finance, information, technology and real estate (Joroff *et al.*, 1993; Barney, 1991).

While it is true that there are many construction stakeholder views of clients, CRE thinking provides for two perspectives on real estate's purpose which is useful for understanding client-ness. This separates CRE, with its use value or demand-for-space perspective, from investment real estate (IRE), which emphasises real estate's wealth aspect (de Jonge *et al.*, 2009). There, the economic value in real estate markets is critical but ultimately real estate wealth does derive from the use of space. This division clarifies the respective interests in real estate. In this conception, CRE encompasses all activities in providing and managing the physical environment for business purposes and embraces facilities management, which has similar concerns. This last conceptualisation is not universal and others offer contrasting views (see for example Jensen *et al.*, 2012).

The CRE perspective founded in CRE's use concepts provides a useful basis for challenging thoughts that CRE and client-ness are incidental to core business. This is because the CRE perspective examines CRE's connections to organisational activities. Variously this has been labelled as 'added value' (Jensen *et al.*, 2012) or CRE 'enabling' competitive advantage (Heywood and Kenley, 2008). A useful model for understanding CRE's centrality to organisations' purposes is Porter's

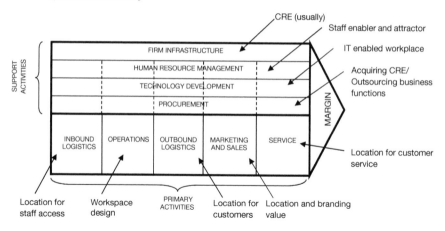

Figure 2.1 CRE pervades organisations' value-creation activities

Source: Adapted from Porter, 1985: 37

value chain model adapted to account for CRE being embedded in organisational processes (Figure 2.1).

The value chain model sees organisations as purposeful, value-creating entities. This model originated in the private sector where the value created is usually economic. The concept is extendable to other value-creating purposes, for example service delivery by public sector organisations. This model's modification is not universally acknowledged in either the CRE literature or the business literature. Nevertheless, we believe that it provides a powerful illustration of the relationship between CRE and value-creation, making it useful in understanding how CRE meets organisational purposes in multiple ways.

Value-creating activities divide into *Primary Activities* (also called 'core business') and secondary *Support Activities* which do not directly create value but are necessary for supporting the organisation's primary activities. CRE is an exception through real estate's value as a real estate asset.

Firm Infrastructure is where CRE is usually located as 'just' the physical properties. The relationship with other support activities has been reconceptualised as corporate infrastructure resources (Materna and Parker, 1998), now also called integrated resource infrastructure solutions (Dunn *et al.*, 2004) where all the support activities are seen as interconnected in supporting organisations achieve their purposes.

This connectedness can illustrate the wider implications of CRE in value-creating activities. For example, CRE's role in *Human Resources* through staff attraction and retention, and enabling their work capacity, has received attention by for example Becker and Pearce (2003) and van der Voordt (2004). *Technology Development* is seen in information technology (IT) becoming ubiquitous in workplaces creating opportunities to trade off real estate and IT costs (Bell, 1994) and increasingly the impact of 'big data'. Workplace designs now planned along activity-based concepts (Appel-Meulenbroek *et al.*, 2011) only work with supporting IT infrastructure. *Procurement* is most usually linked to acquiring operations' resources but also applies to CRE acquired for occupation and the outsourcing of business functions including CREM activities.

Operations (production of goods and services) are affected by the design of the production facility or workspace (Barovick and Steele, 2001). For service organisations much work has been done recently on establishing the link between workplace design and productivity, though this is far from explicitly proven (Haynes 2007). *Logistics (inbound and outbound)* is intrinsically linked to location and is one reason we think a real estate view is useful (location is central to real estate) by impacting on logistics costs in distances from suppliers (including the staff's proximity to place of residence) or to buyers (Gibler, 2006). *Marketing and Sales* is also affected by location through proximity to customers (Fenker, 1999) and physical and virtual modes of contact with customers. CRE also 'says' things about the organisation through physical branding (Appel-Meulenbroek *et al.*, 2010). *Service* locations' proximity for post-sales and customer support affects the customer experience.

The final useful challenge from a CRE perspective, before introducing the CRE models, is the object seen as being of mutual concern to clients and construction. For CRE, founded in real estate concepts, the physical object is more than 'just' buildings. It is a 'property' as a heterogeneous conflation of location, site and the improvements (the building). Real estate concepts, including intangible rights, such as rights of use, rights of entry and the like (Reed, 2007), allows leased real estate based on occupation rights to be valid CRE objects. CRE is also increasingly concerned with 'workplace' bringing together aspects of the building, its fit-out, furniture, and workplace and business support services.

This differentiates CRE from construction where the object of mutual concern is seen as a 'building' (as consistently used by Boyd and Chinyio, 2006). Here, sites are typically seen in terms of ground conditions and access for construction purposes. Designers also tend to see the object as a building or, perhaps, as 'architecture'. Location for them places the building within urban or landscape design contexts.

This distinction is important because procurement and therefore clients and users in a CRE view are about more than just construction and construction services. It also includes existing real estate and its services, workplace and business services, and indeed aspects of the CRE function itself.

Developing the model

We have argued above for the usefulness of a CRE perspective in thinking about clients and also users. Therefore, CRE theory can be deployed to represent clients and users in construction procurement. However, the CRE literature does not yet have a comprehensive account of the CRE organisation in procuring and using construction products and services. What it does have are several models connecting CRE and its management with organisational purpose that provide partial insights into client-ness. Some of these models also include connections with providers of construction products and services. This situation necessitates a new model built from existing CRE theory that not only address the clients and users problem but also advances CRE theory.

This section outlines the method of developing the model of the CRE organisation as a client. A systems-based approach is adopted to model the parts and their relationships. The method is deductive and rationalist based on more than 15 years of CRE research and teaching.

Three CRE aspects with a total of five constituent sub-models provide the basis to the demand-supply model advanced towards the end of the chapter. These are:

- A *supply–demand relationship* where CRE is the demand-side for supply chains of products and services where real estate and construction exist to, in effect, be 'consumed' by clients and users (Sub-model 1).
- An *organisational economic perspective* where CRE's purpose encompasses five roles in CRE organisations' economies that must be addressed and balanced in managing CRE (Sub-model 2).
- A *strategic view* where:
  - ○ CRE has varying levels of 'strategic-ness' with implications for its management – CRE's importance (Sub-model 3);
  - ○ strategic CRE management comprises multiple, interconnected domains of practice – CREM's scope of practice (Sub-model 4); and
  - ○ CRE management evolves through levels of strategic development with implications for the character of CRE management (Sub-model 5).

These five sub-models can be synthesised into a 'super-model' (hereafter just called a model) that describes the client and user system for real estate used for business purposes. The resultant model is complex, which is consistent with the problem in the W118 Research Roadmap (Haugbølle and Boyd, 2013) which also argues that the problem is complex.

As CRE has emerged only recently as a discipline many CRE organisations will not use these concepts as material descriptions of themselves. Nevertheless, from our immersion in the CRE view of organisations we believe that there are useful insights into 'client-ness' and 'user-dom' applicable to all organisations using CRE for value-creation. While it may be a big claim, the model here may go some way to addressing the difficulty noted in the research roadmap of providing an overarching model of the construction clients' world.

Having established the model's development the best way to explain it is to illustrate it with a real case to explain how the sub-models manifest when managing CRE.

Introducing the case example: an Australian bank

This is a case example of a CRE organisation as a revelatory case (Yin, 1994). While not a formal, empirical case study, the understanding of the organisation has developed over 15 years through engagement with its CREM practitioners, attendance at industry presentations, student research projects using public domain information and a doctoral project into one of its workplaces. The case example has been anonymised here.

While arguably any organisation could be an illustration, the bank organisation used here (see Table 2.1) is suitable for several reasons:

- By capitalisation it is one of the world's largest banks.
- It has actively used CRE and CREM as terminologies and practices since at least the mid-1990s.
- It is a complex, space-using entity at the leading edge of CREM practice. Its main CRE challenge, as for all Australian financial services companies, is the 'scale and complexity of the organisation' (Harvey Ross and Marrable, 2015: 8).
- It has a large, diverse property portfolio (Table 2.1) that is for the most part leased to meet its business needs. This requires regular and consistent procurement of CRE from the real estate market.
- Its CREM function operates as a strategic alliance of internal and external entities. Arguably, this constitutes world-leading CREM procurement practice.
- Since the early 2000s it has had a rolling workplace renewal program requiring procurement of new properties and refurbishment of existing properties.

Over the past decade, this bank has been reconfiguring its workspace portfolio to provide flexible workspaces where innovation and knowledge-creation are facilitated. This has resulted in two new 'flagship' head office properties and a refurbished head office property now fully occupied following a base-building refurbishment and the bank's new fit-out. These properties are located in the centre of one Australian state capital, but this workspace concept has also been rolled out to other state capitals' offices. Substantial efforts have gone into change

Table 2.1 Portfolio property types

Workspace	Network	Infrastructure
3 head-office properties totalling 183,000 m² in an Australian state capital city. State branch offices in the other state capitals. Business banking centres distributed throughout metropolitan areas and in regional cities. International offices in New Zealand, Asia & Europe.	Automatic Teller Machines (ATMs) often but not exclusively co-located with other bank properties.	Data centres (mission critical to delivery of services).
Approx. 1,800 retail bank branches		

management with all these projects because, while they are real estate projects, they were also seen as opportunities for organisational development and cultural change. This change is to not only optimise the real estate's use but also to develop behaviours facilitating business outcomes in meeting customers' needs and creating shareholder value.

The network of ATMs and retail branches are generally leased because, in the past, the bank sold and leased back most, if not all, of its retail branch network. It also largely leases its workspace.

The CRE function has an in-house, staff component setting strategy and an external outsourced, property service provider for facilities management, property management and asset management sub-functions. These are co-located within workspace locations. This arrangement is longstanding and has evolved through multiple iterations since the first outsourcing in the mid-to-late 1990s.

The five CRE sub-models and the illustrative case

Sub-model 1: CRE is the real estate economy's demand-side

Based on Heywood and Kenley (2010: 383) this sub-model (Figure 2.2) is the underlying 'scaffolding' for locating the subsequent sub-models. This sub-model has the premise that the ultimate purpose of creating the built environment is its use. This makes the CRE organisation the ultimate demand point in the real estate economy supported by supply chains organised around products (real estate objects) and services. The products could be new construction, existing properties or adaptation of existing properties, and they could be owned or leased. Services include not only those required for construction but also business and property services supporting the CRE's use. These supply chains consist of construction, property and service provider entities.

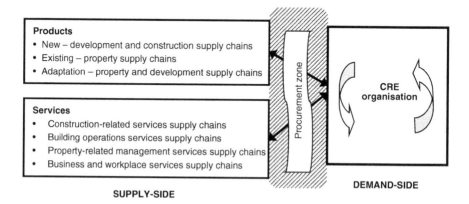

Figure 2.2 A framework for demand–supply entities and relationships

Source: Adapted from Heywood and Kenley, 2010: 383

For the bank, new products included their new head offices, constructed for them by landlord–developers who in turn engaged supply chains of contractors and their sub-contractors, consultants and advisers to deliver the new product. The bank also required project supply chains relating to fit-out design and supply, and change management. Existing property supply chains were found in existing retail branches for which there were no changes. The refurbished head and state office workspaces involved supply chains for products and related services for the adaptation.

The intersection between a CRE demand-side and its supply-side creates a 'procurement zone' between them for the required products and services for creating products, and services in operating and using the products. This zone is also notable for two customer or client concepts. The most frequently met concept is 'customer relationship management', for which many approaches exist. This is a supply-side concept where the supply-side has customers to be managed. The corollary is 'customership' (Kuronen *et al.*, 2011), which fits with a CRE view in that the demand-side CRE organisation bestows its custom on the supply-side as a result of being their customer. The value and importance of this concept is often overlooked because, at present, power tends to reside with the supply-side for several reasons. These include:

- the semantics of 'customer', 'client' or 'tenant' as supply-side concepts with expertise held by the supply-side provider;
- the low weight of numbers in CRE research compared to supply-side research in investment real estate and construction (Heywood and Kenley, 2013);
- real estate theorisation focusing mostly on the supply-side's wealth from real estate investments; and,
- consequently, there is a dominance of 'client' in the discourse about procurement.

Recognising clients' and users' importance shifts the relationship's power dynamic more to the demand-side and opens up the conceptual space for customership to represent this changed power dynamic.

Customership's presence can be seen in the bank's capacity to negotiate its head offices' lease conditions that exceed market norms in terms of a customised building design and integrated fit-out. The bank's quality of lease covenant (largely based on their ability to pay rent) and the extended lease commitment were instrumental in creating these beneficial project and property outcomes.

Sub-model 2: CRE's roles in organisations' economy

The CRE view on the purpose of real estate recognises five roles played by CRE internally in organisations' economy in creating value (Heywood and Kenley, 2013):

- a factor of production;
- a corporate asset;

- an investment;
- a commodity; and
- public infrastructure.

The CRE organisation in making decisions about CRE's purpose needs to balance these roles. For example, current thinking about the role of workplace in a knowledge economy makes the CRE organisation's office spaces an even more important *Factor of production* than it always was. Indeed, many in the CRE and facilities management worlds have taken to calling it a 'strategic resource', for example Joroff *et al.* (1993) and McGregor and Then (1999). Arguably, being a factor of production is the most important role when considering clients and users as this translates reasonably directly into expressed needs for new and altered products. This is clearly seen in new workspace designs in the bank's recent head office properties contributing to cultural transformation.

   Corporate asset and *Investment* roles are largely internal to the CRE organisation but can affect procurement methods. For instance, reducing corporate real estate assets on the balance sheet through leasing property would mean that the CRE organisation may not be a direct client for newly constructed assets with the landlord–developer instead sitting in that position, whereas ownership would make the CRE organisation direct clients. The bank's head offices are an example of this, though in this case the relationship is less pure because of customised design and integrated fit-out. CRE as an investment sees not just owned CRE as investments but also leases and fit-outs as investments because of their medium- to long-term commitment. The *Commodity* role represents the real estate transactions in procurement, that is, real estate capable of development, refurbishment and sales by real estate actors. This was seen in the historical sales of the bank's retail branches. An example of *Public infrastructure* is seen in the externalities from CRE organisations' siting decisions on surrounding localities. These may include negative environmental or positive economic impacts. For example, the first new head office property was instrumental in catalysing the urban renewal of an industrial brownfields location adjacent to the existing city centre.

   It is evident from the case example that internal CRE organisational decision making about CRE roles flowed through to how the organisation acted in procurement and what was required by way of products and services.

Sub-model 3: CRE's level of strategic importance

It is largely self-evident that not every instance of CRE procurement has the same strategic value to the CRE organisation, with consequences for procurement activities capturing the organisation's attention. There are some resulting consequences for procurement interactions. Several CRE strategic classification systems have been proposed: for example, Gibler and Lindholm (2012) suggest different tenure methods to deal with strategic importance – own the long-term required CRE; lease medium-term required space; and rent serviced offices for short-term, project space needs. Should these recommendations be widely adopted then this will change CRE organisation's demands for real estate products and will bring differing commitment levels to projects where the different tenure options apply. Park and Glascock (2010) adopt a resource-based view of firms, in which CRE should be owned only where property's heterogeneity affords resource advantages. Essentially, this only applies to retail CRE where the unique location and possibly store design make it a unique resource conferring competitive advantage. Clearly, if this was adopted as standard CRE practice there would be profound flow-on effects to procuring property and construction.

For the bank, the new head offices were highly strategic, being projects delivering new workspaces and also cultural change. High levels of CRE function attention were evident in the project processes. Interestingly they have opted to lease their long-term, strategic head offices. This may be due to the mechanics of the Australian real estate market and practices favouring leasing, and also the customership noted earlier where a strong demand-side market position was important in securing long-term, customised real estate products held off the balance sheet by a developer–investor.

Sub-model 4: CRE management as multiple, interconnected domains of practice

Proper management of CRE requires multiple domains of practice which reflect the levels and complexity of the internal management task. Using strategic management theory, the Strategic CRE Management Framework (Figure 2.3) organises these into strategic, tactical (here labelled Management and Control) and operational levels. Here, 162 CRE management practices are categorised into domains of practice (Heywood and Kenley, 2008). Varcoe's (2000) model is similar but does not allocate specific practices, though it usefully includes construction, and business and workplace support practices.

Procurement is an expression of the operational level through domains such as location/site selection practices where, for example, in the bank's location choices for its new head offices manifest as new development and construction products. Workplace style practices can flow through to briefed requirements, for instance the bank's environments that support activity-based working. CRE transactions have already been mentioned in own-lease discussions but disposals also create supply-side development and ownership opportunities. CRE project management practices clearly impact on procurement practices, perhaps in the procurement methods adopted, perhaps in the services required to supplement the in-house CRE function's capacity.

The Building Operations' material need for products and services emerges from the higher level, CRE-specific domains of practice. In any project the CRE function must resolve many aspects across these higher-level domains. Changes within, particularly, the strategic domain which connects to value-creating purposes can trigger the need for procurement projects. It is also true that things such as lease expiry in Operations' CRE transactions can similarly act as triggers.

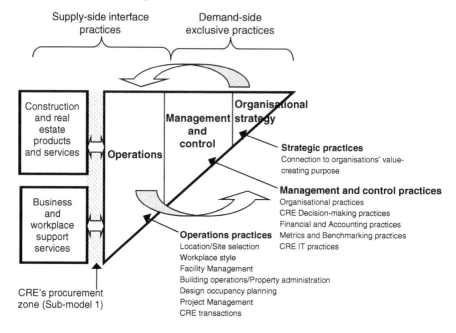

Figure 2.3 The Strategic CRE Management Framework

Source: Adapted after Heywood and Kenley, 2008: 92; Varcoe, 2000: 385

Then location, tenure method, practices for financing the CRE, workplace styles and other domains need resolution in making statements of requirement. Because of this internal CRE organisational practice, we suggest that if the supply-side wishes to better understand 'client-ness' and 'user-dom' rather than just looking at the immediate procurement-oriented operational level it is necessary to look deeper into the CRE management practices that give rise to procurement and use. This sub-model helps to do that.

Sub-model 5: CRE management's level of evolutionary development

This sub-model captures CREM's development of strategic capacity and is based on Joroff *et al.*'s (1993) five evolutionary steps:

- Taskmaster.
- Cost controller.
- Dealmaker.
- Intrapreneur.
- Business strategist.

How CREM deploys the practices noted in Sub-model 4, based on its evolutionary level, has important consequences for the CRE function's defining of itself, for its character and skill sets, and as encountered by the supply-side during procurement. Consequently, the supply-side meets different levels of strategic sophistication, depending on evolutionary levels in different CRE organisations acting as clients. This can also vary over time for individual CRE organisations.

An important implication for understanding clients and users relates to the focus on organisational purpose and understanding CRE's link to the value-creation purpose. At the lower levels, grounded in solid real estate competence in the technical requirements of buildings, properties and their projects, the focus is very heavily on the real estate itself. The organisational strategic purpose, while present, can be obscured by the 'real estate problem'. At the highest level the organisational strategic purpose is paramount with higher levels of senior organisational management involved and decision making based on strategically important information, such as demographic and market forecasts (Joroff *et al.*, 1993). Leading CRE organisations are more likely to be at the business strategist level with everyone else scattered somewhere along the evolutionary continuum.

The bank's earlier sale of retail branches was an expression of cost controlling and deal-making levels through controlling real estate ownership costs and through getting good real estate deals in the disposal. The bank has evolved to the business strategist level where the CRE function is highly attuned to the organisation's strategic objectives and the role of real estate in achieving them.

Conclusion: The proposed CRE client and user model

This chapter adopted a CRE view on the complex 'clients and users' problem to provide a model of the CRE organisation in procurement. That view, internal to the entities (CRE organisations) that are clients and contain users, sees the relationship as a procurement demand–supply one. Here, expressed material procurement requirements emerge from the strategic value of CRE to an organisation's value-creation purpose. The character of the organisation in procurement emerges from the CRE management practices and the degree of evolution towards understanding the strategic connections between CRE, its management and value-creation.

A model is a symbolic representation of reality. Relying on the CRE theory and synthesising the sub-models yields a complex but coherent model of the demand side of procurement (Figure 2.4).

In the model, procurement is seen as a demand–supply relationship based on Sub-model 1 for the supply-side to meet demand for three things:

- New CRE to meet changing business needs (this includes altered CRE).
- Existing CRE to meet continuing needs.
- Services in relation to the above.

This translates into specific requirements for products and services to create, manage and use them. Products include not only new construction but also alterations and existing properties in-use. Because the products are real

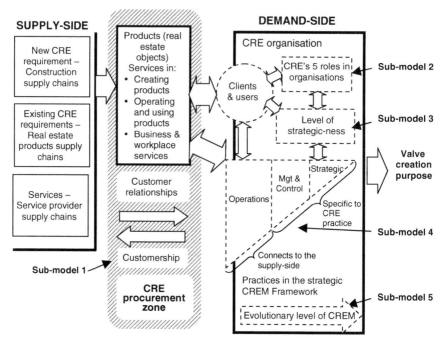

Figure 2.4 The coherent demand-side model in relation to procurement and the supply-side

estate-based they include physical products and leases which are rights of use. Two-way relationship dynamics across the procurement zone are also shown.

The point of intersection between demand and supply is both the material needs for CRE Operations and the intangible phenomenon of being clients and users. The model shows being a client and user as an emergent property (shown literally here) of the CRE organisation's internal dynamics in relation to the CRE and CRE management used in meeting its value-creation purpose.

The roles and strategic-ness of the real estate (Sub-models 2 and 3 respectively) are of concern to strategic CRE management and can emerge in the CRE organisation's procurement routes and their procurement management and effort. This impacts on what it is to be a client and what is asked of the supply-side by way of products and services for use. The CRE organisation's internal strategic and tactical CRE management practices (Sub-model 4) emerge in Operations' external expressions for specific products and services from the supply-side. Practice's degree of evolution (Sub-model 5) also emerges into how the CRE function operates, particularly in whether the task is seen as a real estate one or subtly different, as a business one with consequential effects on its approach to being a client and user.

Using the underlying CRE view, the model opens the 'black box' that is currently construction's clients and users. The CRE view is that internal characteristics of the CRE organisation, its CRE and CRE function represented by the sub-models are instrumental in CRE meeting the organisation's value-creating purpose. They are also instrumental in procurement of construction and property products and services. Because of that instrumentality the model acts as a bridge between the supply-side and organisational purpose that has been identified as necessary to client and user research. The model then serves at least two purposes: how it reveals the supply-side's client and user characteristics; and it describes the CRE world to itself when acting in procurement.

The revelatory case example we provide shows the application to an Australian CRE organisation but because our CRE theorisation draws from the global body of knowledge the model has wide application. By using CRE organisations to symbolise the CRE-using, value-creating entity the model could apply equally to a European or North American multinational corporation, a local government housing agency or housing association, or a small business with a home office. The scale of the task certainly differs but the conceptualisation is common to all. Also, as the thinking that underpins this business, value-oriented model, it also applies to residential real estate but different supply-side supply chains and demand-side motivations would apply. This warrants its own exploration.

Finally, this is a symbolic model. The material reality of CRE in clients and users is most evident in the procurement zone's products and services. Those familiar with the supply-side should relatively readily sense the material organisations that make up supply chains. It is more challenging to identify materiality in our CRE organisation. Some of this may be due to the field's recent emergence. Some may be due to organisations still seeing themselves as 'not in real estate'. Some may be due to not seeing themselves as 'corporate'. And some may be due to the portfolio size as it makes little sense to have a dedicated 'CRE Unit' where there are few properties requiring few decisions. Nevertheless, it is a material reality that organisations require CRE decisions at various times. Here, the CRE function symbolically represents and points to that reality.

The chapter's symbolic model is complex, but so is the problem. The model accounts for both the strategic dimensions to the CRE itself and its management practices. The strategic dimensions are critical for the connection to an organisation's value-creation purposes. Through synthesising five CRE models from CRE theory the internal system of the CRE organisation with its relationships in procurement is described by the model. This uses a different perspective on the problem and opens the hitherto closed black box of client-ness and user-dom. It also points towards a body of knowledge useful to addressing many issues identified as relevant to clients and users research.

References

Appel-Meulenbroek, R., Groenen, P. and Janssen, I. (2011). An end-user's perspective on activity-based office concepts. *Journal of Corporate Real Estate*, Vol. 13, 122–35.

Appel-Meulenbroek, R., Havermans, D., Janssen, I. and Van Kempen, A. (2010). Corporate branding: an exploration of the influence of CRE. *Journal of Corporate Real Estate*, Vol. 12, 47–59.

Barney, J. (1991). Firm resources and sustained competitive advantage. *Journal of Management*, Vol. 17, 99–120.

Barovick, B. and Steele, C. (2001). The location and site selection decision process: Meeting the strategic and tactical needs of the users of corporate real estate. *Journal of Corporate Real Estate*, Vol. 3, 356–62.

Becker, F. and Pearce, A. (2003). Considering corporate real estate and human resource factors in an integrated cost model. *Journal of Corporate Real Estate*, Vol. 5, 221–42.

Bell, M. A. (1994). Hidden occupancy costs. *Harvard Business Review*, Vol. 72, 156.

Boyd, D. and Chinyio, E. (2006). *Understanding the construction client*. Oxford: Blackwell Publishing.

Brown, R. K., Arnold, A. L., Rabianski, J. S., Carn, N. G., Lapides, P. D., Blanchard, S. B. and Rondeau, E. P. (1993). *Managing corporate real estate*. New York: John Wiley & Sons Inc.

CoreNet Global (2007). *Profile of CoreNet Global*. Online. Atlanta, GA: CoreNet Global. Available at: http://www2.corenetglobal.org/home/about_us/index.vsp (Accessed 10 July 2007).

De Jonge, H., Arkesteijn, M. H., Den Heijer, A. C., Vande Putte, H. J. M., De Vries, J. C. and Van Der Zwart, J. (2009). *Corporate real estate management: Designing an accommodation strategy*. Delft: Technical University Delft.

Dunn, D., Ellzey, K., Valenziano, S. F., Materna, R., Kurtz, T. and Zimmerman, C. (2004). *CoRE 2010 integrated resource & infrastructure solutions*. Atlanta, GA: CoreNet Global.

Edwards, V. and Ellison, L. (2003). *Corporate property management: Aligning real estate and business strategy*. Malden, MA: Blackwell Science.

Fenker, R. (1999). The location mystique: Understanding why companies with great sites fail and what to do about it. *Journal of Corporate Real Estate*, Vol. 1, 270–77.

Gibler, K. M. and Lindholm, A.-L. (2012). A test of corporate real estate strategies and operating decisions in support of core business strategies. *Journal of Property Research*, Vol. 29, 25–48.

Gibler, R. R. (2006). Using scorecards to routinely evaluate distribution facility locations. *Journal of Corporate Real Estate*, Vol. 8, 19–26.

Harvey Ross, A. and Marrable, C. (2015). *Emerging practice in corporate real estate: Asia Pacific*. Sydney: Cushman & Wakefield.

Haugbølle, K. and Boyd, D. (2013). *Research roadmap report: Clients and users in construction*. CIB Publication 371. Rotterdam: International Council for Research and Innovation in Building and Construction (CIB).

Haynes, B. P. (2007). Office productivity: A theoretical framework. *Journal of Corporate Real Estate*, Vol. 9, 97–110.

Haynes, B. P. and Nunnington, N. (2010). *Corporate real estate and asset management: Strategy and implementation*. Oxford: EG Books.

Heywood, C. and Kenley, R. (2008). The Sustainable Competitive Advantage Model for corporate real estate. *Journal of Corporate Real Estate*, Vol. 10, 85–109.

Heywood, C. and Kenley, R. (2010). An integrated consumption-based demand and supply framework for corporate real estate. In: Wang, Y., Yang, J., Shen, G. Q. P. & Wong, J. (eds.). *International Conference on Construction and Real Estate Management, 1–3 December 2010 Brisbane*. Brisbane: China Architecture and Building Press, pp. 380–85.

Heywood, C. and Kenley, R. (2013). Five axioms for corporate real estate management: A polemical review of the literature. In: Callagnan, J. (ed.). *19th Pacific Rim Real Estate*

Society (PRRES) Conference. Melbourne: Pacific Rim Real Estate Society. Available at: www.prres.net (Accessed 20 September 2016).

Jensen, P. A., Van Der Voordt, D. J. M., Coenen, C., Von Felten, D., Sarasoja, A.-L., Nielsen, S. B., Riratanaphong, C. and Pfenninger, M. (2012). The concept of added value of FM. In: Jensen, P. A., Van Der Voordt, D. J. M. and Coenen, C. (eds.). *The added value of facilities management: Concepts, findings and perspectives*. Lyngby: Polyteknisk Forlag, pp. 58–74.

Joroff, M., Louragand, M. and Lambert, S. (1993). *Strategic management of the fifth resource: Corporate real estate*. Norcross: The IDRC Foundation.

Kenley, R., Brackertz, N., Fox, S., Heywood, C., Pham, N. and Pontikis, J. (2000). *Unleashing corporate property: Getting ahead of the pack*. Melbourne, Australia: Property Council of Australia and Victorian Department of Infrastructure.

Kuronen, M., Heinonen, J., Heywood, C., Junnila, S., Luoma-Halkola, J. and Majamaa, W. (2011). Customerships in urban housing redevelopment: A case study on retrofitting a suburb. In: Eves, C. (ed.). *17th Pacific Rim Real Estate Society (PRRES) Conference Gold Coast, Qld*. Queensland, Australia: Pacific Rim Real Estate Society. Available at: http://www.prres.net (Accessed 20 September 2016).

McGregor, W. and Then, D. S.-S. (1999). *Facilities management and the business of space*. London: Arnold.

Materna, R. and Parker, J. R. (1998). *Corporate infrastructure resource management: An emerging source of competitive advantage (Research Bulletin no 22)*. Norcross: The IDRC Foundation.

Park, A. and Glascock, J. L. (2010). Corporate real estate and sustainable competitive advantage. *Journal of Real Estate Literature*, Vol. 18, 3–19.

Porter, M. E. (1985). *Competitive advantage*. New York: Free Press.

Reed, R. G. (ed.) (2007). *The valuation of real estate: The Australian edition of the Appraisal of Real Estate* (12th edn). Deakin, ACT: Australian Property Institute.

Rondeau, E. P. (1992). *Principles of corporate real estate*. Atlanta, GA: Edmond P. Rondeau.

Thompson Jr, A. A. and Strickland III, A. J. (2003). *Strategic management: Concepts and cases*. Boston, MA: McGraw Hill Irwin.

Van Der Voordt, T. J. M. (2004). Productivity and employee satisfaction in flexible workplaces. *Journal of Corporate Real Estate*, Vol. 6, 133–48.

Varcoe, B. (2000). Implications for facility management of the changing business climate. *Facilities*, Vol. 18, 383–91.

Yin, R. K. (1994). *Case study research: Design and methods* (2nd edn). Thousand Oaks, CA: Sage Publications.

3 Users in low-energy buildings

Consequences for clients

Frédéric Bougrain and Paula Femenías

Introduction

To mitigate climate change, most European countries have decided to reduce greenhouse gas emissions by a factor of four before 2050. The building industry produces the second largest share of greenhouse gas emissions in terms of energy end usage after transport and has the highest share in the total energy use. Hence, the industry is one of the main target areas. To deal with this challenge, various policies have been launched in Europe, and standards have been issued to reduce energy use in buildings. In countries such as France or Sweden, an increasing number of projects are oriented towards low-energy buildings. While both countries are affected by regional, national and European policies for energy efficiency in construction, there are also differences, which make a comparison interesting.

Sweden has recently witnessed a large increase in low-energy buildings, but policies seem to be lagging behind practice. Conversely, France has been characterised as a country where state intervention is needed to guarantee the wealth and the strength of the economy. This is the case with its energy policy and several laws have been enacted which have contributed to the development of a new stringent thermal regulation.

Sweden has a reputation for applying bottom-up approaches to building design and planning. For example, tenants' participation in public housing development and management has long been the case in Sweden (Bjerken, 1981), but is more scarce in France.

In Sweden, construction clients have the obligation to submit an inspection plan to verify the energy performance of the building. The client is in charge of the control of the maximum energy use defined by the regulation, but the municipality can demand verification and the client needs to set up a monitoring system for energy. Similar obligations do not exist in France.

In both countries, building energy use is defined by the regulation, which includes delivered energy for heating, air conditioning, hot water, operation of building services (pumps, fans, etc.) and other uses in the building (lighting in common areas, elevators, and so on). Similarly, in both France and Sweden, the occupants' electricity use is not included in this definition (household electricity, computers, copy machines, lighting, etc.). However, there are significant differences in national thermal regulation where the design indoor air temperature for heating in Sweden is 21°C while it is 19°C in France. Consequently, users can have a great impact on the total annual energy demand both with respect to building-related energy use such as heating, hot water and occupancy-related electricity use.

This research seeks to investigate the impact of the differences between France and Sweden and to inquire into the role of users on energy consumption. The first objective was to analyse how user behaviours are taken into account in the design and operation of low-energy buildings. The second objective was to assess how users actually influence the energy performance. Indeed, users are frequently considered as responsible for the gap between expected and real energy consumption (Gram-Hanssen, 2013). Some recommendations will be drawn for clients who are responsible for shaping part of the framework of the construction process. Several taxonomies for construction clients have been proposed (Sexton *et al.*, 2008; Tzortzopoulos *et al.*, 2008). The research will only distinguish clients according to their level of experience with construction. The first part examines the role of users and technology related to the operation of the building. The second part defines low-energy buildings in France and Sweden. Then, two case studies focusing on the design, construction and operation of low-energy buildings, one school and one office, are presented. The chapter concludes with lessons learned and practical implications for clients.

This chapter contributes to the research agenda established by the CIB W118 roadmap on 'Clients and users in construction' (Haugbølle and Boyd, 2013: 34) by addressing its desires 'to assess different methods for involving users and stakeholders in decision-making processes on construction as well as operation of built facilities' and also in understanding 'the mechanisms behind successful/ failed projects'.

The role of the user in energy consumption: a literature review

The large influence of user behaviour on the total energy use of buildings has been extensively reported in relation to housing (e.g. Gram-Hanssen, 2013; Ingle *et al*, 2014) but also in relation to offices (Nguyen and Aiello, 2013). In their editorial of the special issue of the journal *Building Research & Information* on 'Housing occupancy feedback: linking behaviours and performance', Stevenson and Leaman (2010: 440) remark: '*even if the building fabric is robust and well insulated with suitable thermal mass, and the home has an efficient energy source, it will still be the inhabitant who ultimately determines how energy efficient a home will be.*'

Brown and Cole (2009) identify two performance gaps in green buildings. The first gap concerns the predicted energy use and the real use of the building (the 'credibility gap'). This is mainly due to the inability of designers to predict how people use the building in practice and this has been reported and analysed extensively (e.g. De Wilde, 2014).

The second gap relates to the difference between assumed and actual comfort. Three potential impact factors contribute to the 'comfort/behavioural performance gap':

- practical/design factors related to the complexity/simplicity of the building regarding readability, accessibility and feed-back from operation;
- behavioural/situational factors concerning the user's experiences as well as knowledge and information provided; and
- social /psychological factors involving the individual sense of responsibility, awareness, expectations and social norms related to use.

The current thinking associated with these three factors is examined through the academic literature.

The usability of the building and the responsiveness of the operator

Low-energy buildings are frequently designed with a combination of passive and active systems. However, it has been stated that it is important to give the occupants control of the advanced and innovative technical systems used to heat, cool and light the building (Hadi and Halfhide, 2010). Catarina and Illouz (2009), who analysed three certified office buildings according to the French environmental assessment scheme HQE, indicate that employees who were not able to monitor the light intensity overrode the time switch. Thus lighting became permanently switched on. Self-regulation makes occupants more sensitive and receptive to environmental messages. For example, Wagner *et al.* (2007) show that the occupants' control of the indoor temperature and the perceived effect of their possible different interventions influence their degree of satisfaction. Many examples show that, when technical solutions generate discomfort, they generate counterproductive reactions in users. Thomsen *et al.* (2005) examined twelve experimental projects (low-energy consumption buildings with solar installations) and reported that in some cases the noise due to the ventilation system induced occupants to block the air supply system.

According to Catarina and Illouz (2009), technology should not be a goal but a tool at the service of easier operation and better performance. '*Keep the design simple. Automation is not always the answer,*' argue Hadi and Halfhide (2010: 63). Comparing the energy consumption of three retrofitted building, Galvin (2014) noticed that complex technologies tend to encounter technical errors. Moreover, due to the complexity, occupants had difficulties in controlling the heating. Moreover, complex technical systems require more maintenance and the associated costs do not always compensate for the energy saved (Branco *et al.*, 2002).

Finally, the quality of service provided to the occupants (appropriate temperature, reliable technical systems or trust in the relationship with the facility manager) strongly influences the behaviours of the occupants (Catarina and Illouz, 2009). Dissatisfied occupants tend to modify the fine tuning of equipment. Thus, the reliability and efficiency of technical systems depends on the behaviour of occupants.

Users' expectations and need of information

The users' understanding of the building and their active involvement in the operation is of high importance to achieve energy goals. Information and instructions to users as well as to the operating personnel is thus paramount. Leaman and Bordass (2007) suggest that users of green buildings are more tolerant to deficiencies and discomfort if they know how the building is supposed to work. There are studies indicating that it is important to repeat energy saving campaigns to inform users and operators as these tend to change over time resulting in increasing energy use.

Several studies report that it is often the case that neither occupants nor operators are trained to understand how buildings work. New systems are rarely presented to the occupants. Thus, due to this lack of communication, occupants tend to judge negatively the installed systems since they have a bad understanding of how to use them (Catarina and Illouz, 2009). In twelve projects examined by Thomsen *et al.* (2005), users complained that they did not receive instructions on how the systems and the building physics work. It was found that moving to a 'green building' was usually a managerial decision and employees were frequently forgotten (Catarina and Illouz, 2009). Once people have moved, it appears that communication should be done in two steps. First, occupants have to be informed about the specificities of the building when they move. Second, a further communication campaign has to be organised at least six months after moving in, when occupants are more aware of their living environment (CSTB, 2007):

> Thus it is important to ensure adequate and appropriate training for all building occupants. It is important that occupants understand not just how to control the building but why they are being asked to control it in a specific way that might be counter-intuitive.
>
> (Hadi and Halfhide, 2010: 63)

However, the interplay between user and building in order to reach energy goals is complex and related to the social context of both production and use (Haugbølle et al., 2015). Paul and Taylor (2008) found that the attitudes and preferences of the users, such as whether they can identify with the concept of the energy-efficient building, will also be important for the overall satisfaction.

Awareness and individual sense of responsibility

Higher energy efficiency in buildings can paradoxically increase the energy use as occupants seek more comfort in the first place. Occupants tend to adjust their behaviour to the efficiency of the building in which they are living. Consequently, energy savings are seldom as large as predicted by the dynamic simulations in the design phase (Gram-Hanssen, 2014). According to Sorrell (2007: VII):

> The potential 'energy savings' from improved energy efficiency are commonly estimated using basic physical principles and engineering models. However, the energy savings that are realised in practice generally fall short of these engineering estimates. One explanation is that improvements in energy efficiency encourage greater use of the services (for example heat or mobility) which energy helps to provide. Behavioural responses such as these have come to be known as the energy efficiency 'rebound effect'.

This rebound effect is also associated with a lack of sense of responsibility. ENERTECH (2012) criticised occupants who open windows during the winter without turning off the heater and who have the temperature above 23°C in order to be able to wear only a t-shirt while indoors both in winter and summer. Galvin (2013) noticed that a relatively small percentage of consumers (23 per cent of households) consumed 52 per cent of the space heating energy. Dard (1986) considered that these behaviours are the result of a hygienist narrative where the opening of windows was considered as very healthy in earlier time. Such social 'practices' that are culturally based are important to address in order to achieve energy saving behaviour (Gram-Hanssen, 2013).

It also appears that during the first year of operation, systems rarely work as expected and need to be adjusted (e.g. new buildings need to dry out, which demands extra heating). Indeed, building systems generally operate according to occupancy assumptions. However, experience shows that the real occupancy in office buildings is much lower than the planned occupancy (Klein *et al.*, 2011). Thus, it requires at least one year in order to optimise the energy systems according to the real occupancy rate and the users' expectations. Moreover, technical defects are frequently more numerous during the first year. This is also due to the lack of involvement of building operators during design and construction phases. Since people moving from a conventional to a green building have usually high expectations of how the building should perform, these technical defects are counterproductive and rapidly demotivate occupants (Catarina and Illouz, 2009); this increases the gap between expected and real energy consumption, especially for innovative low-energy buildings.

Methodological considerations

The main goal for the research has been to learn about how users' behaviour is taken into account during the design of low-energy buildings and in what way the users can influence the results. The phenomenon studied here involves rich information, is context-dependent and has not been explored as a problem. Although the participation of users in planning and design has been put forward as one important foundation for sustainable development of the built environment (Kaatz *et al.*, 2005), the influence on the performance through user involvement in the design of energy efficient buildings seems not to have been studied. The role of higher levels of user participation so as to improve development (and dissemination) of sustainable energy technologies has been studied in single cases of user-led innovation (Ornetzeder and Rohracher, 2006). The problems experienced in the delivery of a complex sustainable building as a result of not having the operation staff and users involved in the design have also been observed (Femenías, 2006). Accordingly, in-depth studies seemed necessary to further explore the problem. A qualitative case study methodology was considered the most appropriate means to get close to the study objects (Flyvbjerg, 2006). The objective was not to predict or to generalise but to describe and to understand.

Considering the aim of getting as much information as possible out of the case studies, the strategic choice of cases was of high importance. As suggested by Flyvbjerg (2006), a critical case involving the phenomenon to be studied will provide more information than a representative or randomly selected case. Cases with outspoken ambitions of producing low-energy buildings were deliberately chosen since they represented 'most likely' outcomes of the case in the two countries. Furthermore, data and feed-back from the building in operation were needed. Thus, the selected buildings needed to have been in operation for at least one to two years. In France, the market of low-energy buildings has emerged in the latter years and the case is consequently a pioneering example. In Sweden, the development of low-energy building has come further, and a case was chosen from a large selection of possible cases representing a client that has taken the step from pilot project to implementing low-energy building as a standard. A third criterion was the willingness of the project owners to share information and data. The selected clients had ambitions to learn from the pioneering projects and invited research to take part of and support their learning process.

Regarding data collection, the case studies were based on documents from the projects, energy data and face-to-face interviews with key actors. The interviews focused on the organisation of the projects, their origins and goals (mainly energy and environmental issues), the characteristics and impacts of main innovative solutions on the operating costs, the competencies of the different stakeholders of the projects, the nature of the contractual agreements, the responsibilities in case of poor performance, the performance of the building in operation and users' involvement during design, construction and operation.

For the French case interviews were carried out between June and September 2013 with representatives from the client (IGN), the architect, the thermal designer, the environmental consultant, the operator and the two people in charge of following the contracts and supervising the operator representing the users of IGN (client and user) and Météo France (user). All respondents received a case report in French and were able to send their comments to the authors. Two external reports completed after 12 and 18 months of operation by a consultant assisting the client during the first two years of operation and were important complementary sources of information. These reports aimed at checking whether energy goals were reached and provided feedback on technical solutions implemented in the building.

Interviews for the Swedish case study were carried out between 2012 and 2014 with: four representatives from the client, one from the architect, one from the HVAC engineer, two representatives from the main contractor and two technicians from the operational unit of the client organisation. In addition, four users were interviewed: the present Headmaster, one teacher and two porters. Besides documents from the process an earlier report focusing on energy measuring were also sources of information.

A first paper based on the same cases has already been developed (Bougrain and Femenías, 2014). This first research examined how energy objectives have modified the balance of power within the construction business system in France and Sweden.

Low-energy buildings

The meaning of the term 'low-energy building' is different from one country to the other and has changed over time with technological progress and the evolution of national thermal requirements. According to the European project Build with CaRe (Tofield, 2012: 6):

> a low-energy new building is a building that is designed to achieve or to come close to the passivhaus standard and one where passivhaus or similar quality processes are followed to ensure that design energy use is realised in practice without compromising occupant comfort and satisfaction.

The European directive on the energy performance of buildings even moved forward and defined a 'nearly zero-energy building', which is a well-insulated and airtight building using a renewable energy supply (European Parliament, 2010). Low-energy buildings are frequently associated with labels such as PASSIVHAUS in Germany, MINERGIE in Switzerland and ENERGY STAR in the United States. Product labelling is one way to proceed further than normal practice and to reduce information asymmetry between owners and buyers/users. The French and Swedish cases illustrate how low-energy standards between countries may differ.

France

Various policies have been launched by the French government to deal with the environmental challenge. Several standards have been set up since the oil crisis of the mid-1970s to reduce energy consumption in buildings. In 2007, a new label was created for low-energy buildings, inspired by the Swiss label MINERGIE. According to this label, the annual primary energy requirement for heating, hot water, lighting, air conditioning and all the pumps required to provide the building energy needs must be lower than 50 kWh/m²/year. To make this more universally relevant, this value is multiplied by a coefficient depending on climate area and altitude, indicating that the consumption value can vary between 40 kWh/m²/year on the Mediterranean coast and 65 kWh/m²/year in the east and north of France. At the end of December 2012, this label became the reference for the new thermal regulation (RT2012). The target for 2020 is 'nearly zero-energy building' to conform to the European Directive. At a local level, regulation is sometimes more stringent. For example, in Paris the maximum primary energy consumption for new-build operations is 50 kWh/m²/year, hence the requirement is stricter than the national standard for this part of the country (65 kWh/m²/year).

Sweden

Sweden has had requirements for the insulation of exterior walls in new constructions since the 1950s. The building energy regulation was strengthened in the wake of the 1970s energy crises but without further substantial updates during several decades. Since 2006, the Swedish energy regulations for new construction have been increasingly strengthened. The last updates came into legal force in March 2015.

At present the building regulation demands a maximum of 70 kWh/m²/year (delivered energy) for multi-residential housing and 60 kWh/m²/year for non-domestic buildings in the southern climate zone IV (the northern climate zones III, II and I add 10, 30 and 45 kWh/m² respectively). If electricity is used as heating source the maximum energy use is 45 to 85 kWh/m²/year. The construction client has the obligation to provide an energy calculation before the construction.

In the past few years, Sweden has seen a rapid development of low-energy buildings. In 2010, 24 per cent of all new multi-residential construction in western Sweden consisted of low-energy buildings, that is, buildings having an energy performance of 25 per cent less energy use than required by the building regulation (Wahlström *et al.*, 2011). This progress has not been pushed by national regulation but rather by commitment among progressive clients and local environmental policy in larger Swedish cities, which have had specific demands for low-energy construction (often around 60 kWh/m²/year, which was 33 per cent lower than that required by the building regulation at the time). Upcoming European demand for near-zero energy building is a contributing factor pushing for innovation in the field of low-energy construction.

The French case

Characteristics of the project

The construction of the new headquarters of IGN (National Geographic Institute) and Météo France (French Meteorological Institute) and a Parisian antenna for SHOM (Hydrographic and Oceanographic Service of the Navy) was decided in 2007 by the Ministry of Ecology, Housing and Transport. The aim was to assemble in one place services formerly located in Paris and its suburbs. The ministry was the client. However, the supervision of the project was delegated to a regional division. Traditional public procurement, which is regulated by law no. 85.704, enacted in July 1985, was applied and led to separate tenders for construction works and maintenance/operation.

When the ministry launched the contest to select the architect, the building was not supposed to be certified for sustainability or as a low-energy building. However, a national multi-party debate on the environmental policy modified the ministry's ambitions. Following the submissions in the procurement process, the client asked the design team to slightly modify the project in order to certify the building according to the French environmental assessment system HQE (High Environmental Quality) and to get the label 'low-energy building'. The HQE certification relates to the operation management system and the environmental quality of the building. Before this project, the client (the ministry) did not have any experience with low-energy buildings.

The cost of the building works reached EUR 30 million for a gross floor area of 14,900 m² (and 180 parking places). The building hosts about 620 people.

The integration of user needs at the design phase

The users were not involved during the design process. The people in charge of the facility management at Météo France and IGN discussed the access control, electrical current and transfer of professional equipment together with the architect at the design stage. They also sent proposals to the architect regarding furniture and materials in the entrance hall. However, the architect did not take their suggestions into account.

Several technical and architectural decisions have come to have a strong impact on the use. For example, the environmental consultant to the client optimised the size of the windows in order to find the right balance between daylight and thermal supply. Consequently, all offices and all meeting rooms benefit from natural lighting. Moreover, there are sensors that monitor the light intensity in order to control the artificial lighting.

Some proposed solutions were refused by the client who thought that these could lead to dissatisfaction among the users; for example, the environmental consultant wanted to promote natural ventilation, but the client considered that such a solution would not have been acceptable in the summer months. Moreover, a natural ventilation system would have required the use of a mechanical system to open the window for night ventilation, which would have increased construction and operation costs.

The role of users during the operation of the building

A private facility manager/operator maintains the building and monitors the energy use. This facility manager was involved after the delivery of the building. The contract was signed for one year and it could be extended three times. The client wanted to launch a tender for an energy-saving performance contract. However, IGN, which monitors the facility manager, considered that it was too early since the building was new and did not have any record for energy consumption.

Users report problematic issues to an onsite helpdesk, which is the interface between the users and the facility manager. It also allows for records of response times and data on the pattern of operational problems. Once the users inform the helpdesk, the facility manager has to satisfy the demand within a certain time limit, which depends on the significance of the part of the building concerned. This approach was new for people working at Météo France who had never recorded and formalised maintenance activities before moving. Consequently, a specific organisation was developed to comply with the service level requirements. However, it has been found that urgent matters are sometimes recorded after being solved in order to keep a record of the problem and its solution.

Occupancy and energy consumption

After one year of operation, energy consumptions were much higher than expected (see Table 3.1). This was due to the dysfunctions of some technical systems (e.g. the geothermal) and a gap between the theoretical energy value and the real use of the building.

The gap between theory and practice is in part explained by failed specifications of the building. Two examples illustrate this. First, one floor dedicated to Météo France was in operation 24 hours per day 7 days a week – that is, 168 operating hours per week – while the air processing system was supposed to work only five days a week from 8am to 5pm equal to 45 operating hours per week. Since the seven floors are interdependent, the air processing system had to be in operation 168 hours per week. Second, the entrance hall was supposed to be heated to 17°C. This temperature is fine for employees who cross the hall but not for the receptionists who work there. The architect had refused to build a specific 'box'

Table 3.1 Comparison between energy objectives and consumptions in operation

Use	Target	Actual consumption May 2012– April 2013	Actual consumption Nov. 2012– Oct. 2013	Gap for the last period
	$kWh/year/m^2$	$kWh/year/m^2$	$kWh/year/m^2$	%
Heating	3.78	21.95	24.81	+555 %
Air conditioning	4.27	19.26	20.72	+384 %
Hot water	0.19	–	–	–
Lighting, office automation	38.48	49.88	48.98	+27 %
Ventilation and auxiliaries	4.48	11.83	11.72	+161 %
Total without PV	51.20	102.93	106.24	**+107 %**
Photovoltaic	1.20	0.81	1.05	−11 %
Total	50	102.12	105.19	**+110 %**

Source: Adapted from BEHI, 2013: 4

for receptionists since this would have interfered with the aesthetic visions of the architect. Instead, heaters were installed above the receptionists and also at the sides of the entrance itself. As a result, most of the energy used to heat the building was used in the entrance hall.

Users' satisfaction and communication during operation

During the first year, it was found that the temperatures were the same from one floor to another. According to a report ordered by the client, temperatures on 22 April at noon ranged from 21.6°C on the ground floor to 24°C on the seventh floor. At this time, the centralised control station was not properly adjusted. The responsiveness of the building operator was limited because when he was appointed the handover of the centralised control station had not been accepted by the client. Thus, for about five months, the operator did not modify the parameters, in order to avoid breaching liability.

A post occupancy survey was carried out by a student in June 2013. About half of the occupants answered (306 respondents out of 620). Among the 306 respondents, 172 worked at Météo France, 126 at IGN and 8 at SHOM. According to the survey, about 42 per cent of the respondents experienced cold during the first year. At the same time, 28 per cent reported that they opened their window during the winter, while 9 per cent (27 people) indicated that they brought their own heater to their office as temperatures on winter mornings could be below 19°C. According to the logistics manager at IGN, this behaviour disturbed the balance of the heating system. Users were allowed to increase/decrease the temperature of their office by only 1°C. Thus, the control system was experienced as not being reactive enough, which was also explained by the high thermal inertia of the building. Among the 108 respondents who tried to modify the temperature of their office, only 21 per cent were satisfied. According to the operator, the high thermal inertia made the management of the building complicated during spring and autumn. During these seasons, there is a need to heat the building early in the morning, which might then lead to over-heating in the afternoon.

According to the survey, the users appreciated the natural lighting of the meeting rooms and their office. However, they were very critical about their inability to modify the intensity of the light of their office, which was monitored by sensors. During mid-seasons, despite window shutters, some users had been disturbed by the sun reflecting onto their computer screens. According to the architect, this problem could have been avoided if the users had followed her layout for each office.

Météo France provided their employees with a welcome booklet. Moreover, during the first three months, posters were displayed in the corridors in order to explain the principles of the building in operation (e.g. lighting controlled with sensors, the role of the thermostat in every office, etc.).

IGN developed a similar approach. A message was sent to the users about the building in operation. It mainly concerned the heating and lighting systems (the lighting intensity varies with the external brightness). Nevertheless, according to the satisfaction survey, this information was not sufficient: 84 per cent of the interviewed occupants (257 respondents out of 306) claimed that they had not been trained and that they had not received any explanatory note on how, and why, to control the building in a specific way.

Conclusion from the French case

The French case illustrates the 'credibility gap' – a mismatch between prediction and measurements – put forward by Brown and Cole (2009). It results from poor links between users, client and the design team. During the operation of the building this situation was not compensated by the relationship between the users and facility managers and a communication plan that would modify the behaviour of the occupants.

The Swedish case

Characteristics of the project

The Brottkärr project is the addition of a new annex and the refurbishment of an existing building to host a pre-school and a school for up to 300 children under 12 years old. The total area is 3,600 m². The building has been in use since August 2011. The building is heated with a geothermal system and a balanced ventilation system with heat recovery and CO_2 control. The building design is also adapted to low-energy use with temperature zones, a compact form and thermal inertia. A two-year guaranteed follow-up was completed in August 2013.

The client, a municipal agency, owns and operates facilities for the city of Gothenburg. Their property includes pre-schools, schools, housing for elderly and housing for people with special needs. Since 2011, the agency has an owner directive to construct only low-energy buildings, defined as 45 kWh/m²/year in delivered energy. By the time of the design this was 45 per cent lower than the building regulation, which dictates 80kWh/m²/year for non-domestic buildings. During 2013, the agency delivered more than twenty low-energy buildings, mainly schools and nurseries.

The integration of user needs at the design phase

The architect and the engineers appointed for the design did not have any problems in theoretically achieving the energy target. During the design process, the client demanded an energy balance calculation on several occasions. However, depending on the method used the consultants reached different results (see Table 3.2). Since 2010 the client has been using two parallel methods for energy balance calculations: method 1 is based on the client's own method and data, and method 2 follows the method proposed by the building regulations.

The Headmaster and some employees from the school, representing the users, were invited to take part in the design process, although only if their working

Table 3.2 Calculated and measured values for Brottkärr School

	Target	Client's own method	Method according to building regulation	Real energy use
		2010	2010	2013
	kWh/m²/year	kWh/m²/year	kWh/m²/year	kWh/m²/year
Heating need[1]	–	104	103	No data
Heat gain/heat recovery[2]	–	-67	-74	No data
Total heating need	**19**	**37**	**29**	**25.5**
Electricity for operation of building	16	6	17[3]	10 (19)[4]
Delivered energy from heat pump			-34	No data
Specific energy need	45	43	12	35.5 (44.5)[4]
Electricity for users' activities	–	9	9	21[5]
Total electricity need	–	**15**	**26**	**40**

Source: Gothenburg City Lokalförvaltningen (2010: 2) and personal communication with Gothenburg City Lokalförvaltningen.

Notes

1 Including transmission and ventilation losses, losses in systems, and hot water.

2 Heat gain from insulation, internal heat sources and heat recovery.

3 Including 11 kWh/m²/year to operate the heat pump.

4 Including heat pump.

5 Deduced, not measured.

tasks permitted. There was no specific time allocated for the future users to take an active part in the project. During the construction phase, the Headmaster was changed twice. This represents a discontinuity for the user involvement in the delivery of the new facility. Furthermore, there was little documentation from the earlier stages of the process that could be communicated to the new participants. In fact, the involvement of the new Headmaster, who entered late in the process, mainly consisted of getting continuous information about the progress of the construction work. On the whole, the users found their involvement to be limited to the approval of the early design and drawings.

The role of users during the operation of the building

In order to reach an optimal performance of the low-energy system, automated systems were installed which were not dependent on active user involvement. There were sensors to regulate lighting, heating and ventilation, and, according to a representative from the project team, users and the operational staff were told not to alter these.

The operating technicians experienced initial problems with the building, and many adjustments had to be made after handover. The technicians said that they had not been informed about this being a low-energy building, and that they had not received any special training or information about the systems. However, such low-energy buildings do not usually involve technologies that are completely unknown to the technicians. A more general problem lies in the small amount of time that the technicians were given to operate and manage buildings. The technician responsible for the operation of the school at handover changed his employment due to a very high workload as he alone was responsible for 170 facilities and operated more than 300,000 m².

One technician working at Gothenburg's local premises administration said that the project division at the client focused on new solutions without considering the effect on the operation. He would have liked to be more involved in the early stages of the design process, but normally there was no time for this. To quote the technician: '*As project leader, you want to do something that catches the eye, test new stuff. And the ones who are affected are the operating technicians. It is difficult to get balance in the technical systems.*'

Occupancy and energy consumption

The energy use of the school building was higher than expected. The client, the municipal agency had the same experience from several of their low-energy facilities. Hence, the client initiated a project, with Brottkärr School as a case, to develop methods for continuous control of energy targets through the entire process, from design to use.

Moreover, the client was faced with a lack of available data on user behaviour for schools and other kinds of special buildings. The agency is now attempting to define such data and has also developed its own calculation method, which they find better reflects reality than the calculation method proposed by the Swedish building regulation.

The two-year guarantee inspection indicated that the building did not function as planned. Consequently, the client ordered a follow-up by an external consultant, which also served a more general purpose of advancing their knowledge on low-energy buildings. There have been some problems with the monitoring of energy use and not all separate energy flows have been measured (Table 3.2). Preliminary results from the follow-up study indicated that the energy gain from internal heat sources and from passive solar energy were overestimated. In addition, the indoor temperature was suspected to be higher than the calculated 20°C. Furthermore, electricity use for fans, pumps, etc., exceeded that calculated and the operating time of the fans were underestimated. The users were suspected of interfering with the system. The entrance doors were heavy in order to shut quickly when opened; however, users tended to leave these doors open during recreation as the original automatic door openers were quickly worn out.

User satisfaction and communication during operation

Teaching staff was rather disappointed with the building, especially with the heating and ventilation systems, but also with some functional aspects. Their disappointment can be traced back to their limited influence on the design but also on communication problems with the facility manager.

In a similar way to the operation technicians, users reported that they were not informed about the low-energy measures or in which way this would affect the use of the building. The users were dissatisfied with the fact that they could not adjust the heating or the ventilation. During the night and at weekends, the heating and ventilation were automatically lowered to a minimum, resulting in temperatures of 14°C when school started on Monday morning and poor air quality.

The extreme air-tightness also resulted in over-heating, especially on sunny days. The ventilation system had been pre-installed to start when temperatures rose, and the users experienced temperatures of 23–26°C in the classrooms, which they consider too high. Sun-shading that should have been installed had been omitted, and the blinds quickly became worn and deteriorated due to frequent unprogrammed use.

Conclusions from the Swedish case

In the Swedish case, it is the same municipal agency acting as client, building operator and facilities manager. Yet there were problems linking the design to operation and use. The agency has reacted to complaints among the technicians and, since the case study, has employed more staff for operating their facilities.

The Swedish case shows problems with integrating users and behaviour in the design phase. There was a lack of processed information on user habits that could be used as data in calculations. Although the users were involved in the design process, their participation was strictly limited due to lack of time, knowledge and discontinuity in the process as the user representatives changed during the project.

Discussion and lessons learned

The comparison between the French and the Swedish cases support most issues identified by Brown and Cole (2009). The two cases illustrate the credibility gap (the mismatch between predicted energy value and the real performance of the buildings); in both cases, it resulted from the inability of the designers to integrate the future behaviour of the future occupants. However, even when they took part in meetings, occupants were unable to anticipate their future working environment. Thus, there is a need in such situations to rely on professional clients or to involve intermediaries representing the occupants.

The gap in experience between the French and the Swedish clients speaks in favour of agencies that procure, own and operate public buildings, are in charge of planning and managing large property projects, and act as professional clients.

Inexperienced purchasers of construction services with limited expertise in procurement process, such as the French client, are frequently the cause of a building's poor performance.

'Lead clients' are needed to modify the frontier between design, construction, operation and use stages. They need to be professional and able to clarify and interpret users' needs, establish a hierarchy among them and produce an occupier brief which compromises between users' expectations and energy constraints. In this case, the involvement of users at the design stage may not have been required. However, the client needs to involve operators at this level since clients cannot speak for operators.

Unsophisticated clients need to develop stronger relationships with both future occupants and operators. To compensate for their lack of experience, they need to find intermediaries (architects, designers, consultants, etc.) who can identify and interpret users' needs and integrate the requirements of the operators at the design stage. The aim would be to co-produce the occupier brief, which up to this point has only focused on design issues, so as to describe the required functionality of the facility in technical and 'usable' terms, with the identified stakeholders. Table 3.3 summarises these relationships between clients and the other stakeholders of a building project at the design and operation stages.

The cases show the growing use of automatic systems regulating heating, hot water, air-conditioning and lighting. These are meant to ensure that an optimal performance of the low-energy system is achieved. However, this approach reduces the users' ability to dramatically influence the energy use. This situation could be counterbalanced by a closer relationship between the occupants and the operator. In the French case, users report problems to an onsite helpdesk which is the interface between the users and the facility manager. It also allows for records of response times and data on the pattern of operational errors. However, the responsiveness of the operator regarding complaints about the temperatures was poor. Similarly, in the Swedish case, users were dissatisfied since they were not able to adjust the heating or the ventilation. The dry in-coming air even led to health issues. Moreover, users reported communication problems with the facility managers. In both cases, the operator was more a contractor than a partner. In France, they were selected after the handover. Moreover, they signed a one-year

Table 3.3 Impact of client professionalism on the relationship with occupants and operators at different stages of a project

	Design stage		Operation stage	
	Professional client	*Unsophisticated client*	*Professional client*	*Unsophisticated client*
Occupants/ intermediaries representing occupants	Internal interpretation of users' needs	Co-production	Information/ participation	Information/ participation
Operators	Consultation/ cooperation	Co-production	Cooperation	Co-production

contract which could only be extended three times. In Sweden, the technician responsible for the operation of the school belonged to the same organisation as the client. However, he did not feel engaged with the project as he was also solely responsible for operating a total of 170 facilities. Due to his high work load, he finally resigned.

The second lesson would thus be that the development of automatic systems needs to be counterbalanced by the development of new relationships between occupants and operators. There is a need to consider the operator as a partner not only for the client but also for the occupants. Mid- to long-term contracts are necessary to provide operators with incentives to regulate the building for the long run and to develop relationships based on trust with the users.

The third issue has to do with the need to inform the users about the buildings systems. The literature emphasises the importance of informing and activating the users in order to counteract their potential disruptive behaviour. Yet, in both cases, occupants received little information about their building. When it was done, users were not familiar with their work environment and they had no questions about surrounding systems and equipment. This situation could not raise the awareness and the sense of responsibility of users (one of the issues identified by Brown and Cole, 2009). There is even a risk in the long run of developing a counterproductive situation that would demotivate occupants and reinforce the rebound effect.

References

BEHI (2013). *Suivi d'exploitation: Visite du 10/12/13 IGN–Météo France [In French: Operation follow-up: Visit of 10 December 2013 IGN–Météo France]*. Ramonville-St-Agne: BEHI.

Bjerken, M. (1981). Public participation in Sweden. *Town Planning Review*, Vol. 52 (3), 280–85.

Bougrain, F. and Femenías, P. (2014). Delivering and operating low-energy buildings in France and Sweden. In: Jensen, P. A. (Ed.) (2014). *Proceedings of CIB Facilities Management Conference: Using Facilities in an Open World – Creating Value for All Stakeholders. 21–23 May 2014*. Lyngby: Technical University of Denmark.

Branco, G., Lachal, B., Gallinelli, P., Gonzales, D. and Weber, W. (2002). *Analyse thermique de la Cité Solaire à Plan-les-Ouates (Genève) [In French: Thermal analysis of Cité Solaire at Plan-les-Ouates (Geneva)], rapport final, Rapports de recherche du CUEPE no 1*. Geneva: CUEPE.

Brown, Z. and Cole, R. J. (2009). Influence of occupants' knowledge on comfort expectations and behavior, *Building Research & Information*, Vol. 37 (3), 227–45.

Catarina, O. and Illouz, S. (2009). *Retour d'expérience de bâtiments de bureaux certifiés HQE: dynamiser l'efficacité énergétique des gestionnaires de patrimoine du secteur privé [In French: Feedback from office buildings benefiting from HQE certification]*. Paris: CSTB, ICADE, ADEME.

CSTB (2007). *Comparaison internationale bâtiment et énergie, Rapport final, Programme PREBAT, ADEME, CSTB, PUCA [In French: International comparison of building and energy]*. Paris: CSTB.

Dard, P. (1986). *Quand l'énergie se domestique ...Observations sur dix ans d'expériences et d'innovations thermiques dans l'habitat, Recherches, Plan Construction [in French: When energy is domesticated ...Observations over ten years of experimental projects and thermal innovations in housing]* . Paris: Plan Construction.

De Wilde, P. (2014). The gap between predicted and measured energy performance of buildings: A framework for investigation, *Automation in Construction*, Vol. 41, 40–49.

ENERTECH (2012). *Grenoble – ZAC de Bonne: Evaluation par mesure des performances énergétiques des 8 bâtiments construits dans le cadre du programme européen Concerto – Rapport de Synthèse [In French: Grenoble – ZAC de Bonne: Evaluation of the energy performance of eight buildings developed in the framework of the European project Concerto – Final report]* . Félines sur Rimandoule: ENERTECH.

European Parliament (2010). *Directive 2010/31/EU of the European Parliament and of the Council of 19 May 2010 on the energy performance of buildings*. Brussels: Official Journal of the European Union.

Femenías P. (2006). Experiences from Universeum Science Centre in Sweden: Focus on the delivery phase a weak link between the project and the building in use. In: *Proceedings from the 23rd International conference on Passive and Low Energy Architecture, Geneva 6–9 September*. Geneva University: CUEPE, 939–44.

Flyvbjerg, B. (2006). Five misunderstandings about case-study research. *Qualitative Inquiry*, Vol. 12 (2), 219–45.

Galvin, R. (2013). Targeting 'behavers' rather than behaviours: A 'subject-oriented' approach for reducing space heating rebound effects in low energy dwellings, *Energy and Buildings*, Vol. 67, 596–607.

Galvin, R. (2014). Making the 'rebound effect' more useful for performance evaluation of thermal retrofits of existing homes: Defining the 'energy savings deficit' and the 'energy performance gap', *Energy and Buildings*, Vol. 69, 515–24.

Gothenburg City Lokalförvaltningen (2010). *Energianalys. Brottkärrskolan Bedömning av byggnadens energianvändning. CeAK VVS konsult* (Energy analysis. Brottkärrskolan. Assessment of the building's energy use. CeAK VVS consult), 15 February. Internal document.

Gram-Hanssen, K. (2013). Efficient technologies or user behaviour: Which is the more important when reducing household's energy consumption? *Energy Efficiency*, Vol. 6 (3), 447–57.

Gram-Hanssen, K. (2014). Retrofitting owner-occupied housing: remember the people, *Building Research & Information*, Vol. 41 (4), 393–7.

Hadi, M. and Halfhide, C. (2010). *Are building users sabotaging the move to low-carbon buildings? BRE Trust Review 2009*. Watford: BRE Trust.

Haugbølle, K. and Boyd, D. (2013). *Clients and users in construction: Research roadmap report*. CIB Publication 371. Rotterdam: CIB.

Haugbølle, K., Forman, M. and Bougrain, F. (2015). Clients shaping construction innovation. In: Ørstavik, F., Dainty, A. R. J. and Abbott, C. (Eds.). *Construction Innovation*. London, UK: Wiley Blackwell, 119–134.

Ingle, A., Moezzi, M., Lutzenhiser, L. and Diamond, R. (2014). Better home energy audit modelling: incorporating inhabitant behaviours, *Building Research & Information*, Vol. 42 (4), 409–21.

Kaatz, E., Root, D. and Bowen, P. (2005). Broadening project participation through a modified building sustainability assessment. *Building Research & Information*, Vol. 33 (5), 441–54.

Klein, L., Kavulya, G., Jazizadeh, F., Kwak, J.-Y., Becerik-Gerber,B., Varakantham, P. and Tambe, M. (2011). Towards optimization of building energy and occupant comfort using multi-agent simulation. In: *International Symposium on Automation and Robotics in Construction 2011*. Seoul: IAARC. Available at: http://www.iaarc.org/publications/fulltext/S07-5.pdf (Accessed 29 February 2016).

Leaman, A. and Bordass, B. (2007). Are users more tolerant of 'green' buildings? *Building Research & Information*, Vol. 35, 662–73.

Nguyen, T. A. and Aiello, M. (2013). Energy intelligent buildings based on user activity: A survey. *Energy and Buildings*, Vol. 56, 244–57.

Ornetzeder, M. and Rohracher, H. (2006). User–led innovations and participation processes: lessons from sustainable energy technologies, *Energy Policy*, Vol. 34, 138–50.

Paul, W. L. and Taylor, P. A. (2008). A comparison of occupant comfort and satisfaction between a green building and a conventional building, *Building and Environment*, Vol. 43, 1858–70.

Sexton, M., Abbott, C. and Lu, S.-L. (2008). Challenging the illusion of the all-powerful clients' role in driving innovation. In: Brandon, P. and Lu, S.-L. (Eds.). *Clients Driving Innovation*. Oxford: Wiley-Blackwell, pp. 43–8.

Sorrell, S. (2007). *The rebound effect: An assessment of the evidence for economy-wide energy savings from improved energy efficiency.* London: UK Energy Research Centre. Available at: http://www.ukerc.ac.uk/programmes/technology-and-policy-assessment/the-rebound-effect-report.html (Accessed 29 February 2016).

Stevenson, F. and Leaman, A. (2010). Evaluating housing performance in relation to human behaviour: new challenges. *Building Research & Information*, Vol. 38 (5), 437–41.

Thomsen, K. E., Schultz, J. M. and Poel, B. (2005). Measured performance of 12 demonstration projects: IEA Task 13 advanced solar low energy buildings. *Energy and Buildings*, Vol. 37, 111–19.

Tofield, B. (2012). *Delivering a low-energy building: Making quality commonplace*, Build with CaRe. Norwich: Adapt Low Carbon Group. Available at: http://archive.northsearegion.eu/files/repository/20140331180312_BuildwithCaReResearchReport-DeliveringaLow-EnergyBuildingOct2012.pdf (Accessed 29 February 2016).

Tzortzopoulos, P., Kagioglou, M. and Treadaway, K. (2008). A proposed taxonomy for construction clients. In: Brandon, P. and Lu, S.-L. Lu (Eds.). *Clients Driving Innovation*. Oxford: Wiley-Blackwell, 58–68.

Wagner, A., Gossauer, E., Moosmann, C., Gropp, Th. and Leonhart, R. (2007). Thermal comfort and workplace occupant satisfaction: Results of field studies in German low energy office buildings. *Energy and Buildings*, Vol. 39, 758–69.

Wahlström, Å., Jagemar, L., Filipsson, P. and Heincke, C. (2011). *Marknadsöversikt av uppförda lågenergibyggnader. [In Swedish: Market review of low-energy buildings]*. Lågan report 2011:1. Gothenburg: Lågan.

4 An ethical foundation for health and safety

Philip McAleenan and Ciaran McAleenan

Introduction

The construction industry client has a central responsibility to business progression and societal development. Accordingly the client is positioned at the centre of several ethical relationships: they procure construction to fulfil their needs; their business interacts with society and works with its users and customers; and their projects integrate into the social and natural environments (Haugbølle and Boyd, 2013). Such responsibility is the ethical dimension that underpins governance, agency and the management of construction projects. Globally the industry has a reputation that generally falls well below what can reasonably be called ethical, being as it is an environment where competition between contractors leads to a low price mentality, fierce competition and paper-thin margins, and thus to quality and safety reductions to cut costs and save time. Abdul-Rahman *et al.* (2007, 2010) summarised research in Australia and South Africa that identified several unethical conducts and ethical dilemmas in the construction industry such as: corruption, negligence, bribery, conflict of interest, bid cutting, under bidding, collusive tendering, cover pricing, front-loading, bid shopping, withdrawal of tender and payment game.

Following the 2008 global recession, construction suffered badly as a result of the cancellation or suspension of major infrastructure projects, negative equity housing and the banks calling in of loans from small contractors. Competition and reduced margins remain a central feature of a landscape in which the ILO *et al.* (2008) predicted that such a climate would have a direct negative impact on the safety and welfare of workers as clients sought to minimise losses. ILO (2012) states that there has been and continues to be reduced market incentives for clients' enterprises to expend finances on socially desirable levels of investment for better working conditions as a result of the externalisation of costs of injury and ill health onto workers and society through:

- publicly funded medical programmes and insurances picking up on some enterprise costs;
- workers picking up the entirety of the quality of life costs; and
- the public purse picking up on social security cost to the worker and their family (ILO, 2012).

This chapter explores the consequences of this with respect to the construction environment, focusing on the social costs of the situation where the consequences are often externalised from the client. When Egan (1998: 4) suggested that *'If the industry is to achieve its full potential, substantial changes in its culture and structure are also required to support improvement ...'*, he was emphasising the role of culture set within the greater ethical and agency debate of practice and the way moral and ethical norms are developed. This allows the ethical foundation and its main themes – reasoning, agency, sustainability and competence – to be explained and their implications for practice discussed. The ethics foundation forms the basis of the Organisation Cultural Maturity Index, a diagnostic metric aimed at ascertaining the level of workplace cultural maturity within the construction industry and providing the client with a development plan for corporate and societal sustainability (McAleenan, P., 2016).

Global costs of occupational safety and health failures

The global costs of workplace accidents and ill health are substantial; the best available data available from the International Labour Organization (ILO *et al.*, 2008) shows the number of annual fatalities in the workplace to be approximately 2.34 million with a further 317 million workers experiencing injuries following accidents at work and 160 million people newly suffering ill health from workplace illnesses and diseases. The figures are not evenly distributed across the world nor across social sectors, and developing countries suffer disproportionately where larger numbers of the population are engaged in hazardous activities such as mining, farming and fishing, while poorer people, often women, children and migrant workers, are least protected. The cost of accidents worldwide have been calculated by Dorman (2012) at 4 per cent of gross domestic product (GDP), which is approximately USD 3,399 billion derived from 2012 purchasing power parity calculations, the equivalent of a sixth place in the global ranking of GDP (Figure 4.1).

As noted, accidents and work-related ill health are costly in human terms as well as in financial terms. In Great Britain over half the costs (Figure 4.2) are borne by the injured person and included in this is an accounting for non-financial human costs, i.e. the monetary value that individuals would be willing to pay to avoid risk of death or ill health or injury, a figure that amounts to substantially more than half of the total costs.

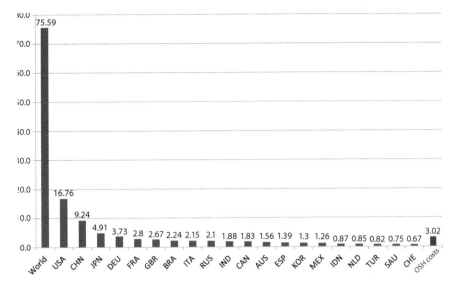

Figure 4.1 Cost of workplace injuries as a ranking in global GDP (USD trillion), 2012, top 20 economies illustrated, global GDP in total USD 75.6 trillion

Source: Dorman, 2012

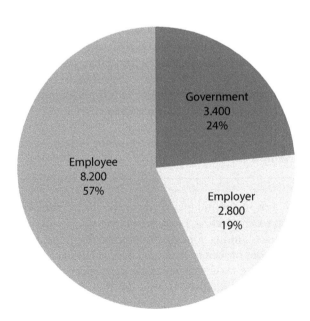

Figure 4.2 Costs of workplace injuries and work-related ill health in Great Britain, 2013/14 (GBP million)

Source: Health and Safety Executive, 2015: 11

End-user impact on client behaviour

The International Labour Organization data relates, on the surface, to the relationship between employer and employee and the human and economic costs that may be externalised to the social purse.

In the case of Qatar and the World Cup the prospective end user of the product can impact positively on how the process of construction should take place and on the conditions in which construction workers should be expected to work (Box 4.1). Following the Rana Plaza building collapse in 2013 in which 1,129 workers died, the primary customer for the garments in Europe and the USA pressurised the retail outlets to develop ethical policies in regard to their supply chain and a new accord for fire and building safety in Bangladesh was created (Bangladesh RMG, 2013).

Box 4.1

FIFA World Cup 2022 in Qatar

In Qatar, there are approximately 1.2 million migrant workers, many working on the 2022 World Cup construction projects. In the four years since 2010 the Indian government has confirmed that 974 Indian migrant workers have died in workplace accidents; and 383 Nepalese migrants have also died as a result of workplace accidents since 2012 (Gibson, 2014). These figures are controversial, not solely for being particularly high for construction, but also because the client, the Qatar 2022 Supreme Committee, represents the government and FIFA, the governing body of international football, responsible for the organisation of the World Cup, who are accountable to millions of sports fans throughout the world.

In 2012 Qatari response to workplace accidents was to publish, through the Qatari Foundation, *Mandatory Standards for Migrant Workers Welfare*, which included a section on equal and humane treatment of all workers (Qatar Foundation, 2012).

The competitiveness in the industry leads to the ills of under-bidding, bribery, collusive bidding, etc., as well as extensive human rights violations, particularly the rights to adequate housing, forced evictions and displacement of marginalised and vulnerable sections of society (Rolnik, 2009). The client has a central role to play in determining the extent to which such practices should be permitted or curtailed. Setting the standards and ensuring that they are implemented are necessarily combined. The key components are a clear policy statement, a transparent organisation with mechanisms to deliver excellence in procurement, design and delivery, a process to change practices so that health and safety really matter and through this to become a beacon of excellence. Though establishing an ethical commitment to workers' health and wellbeing should be a sufficient guide, the International Labour Organization (ILO, 2012) argues that occupational safety and health (OSH) improvements are evidentially economically beneficial and therefore, though ethically redundant, provide a motivation for clients to demand good safety and health practices. Nevertheless the client–contractor relationship remains one wherein the client retains the capacity to influence what projects are to be funded and how they are to be carried out, if not in technical terms then certainly in ethical terms where they set guiding rules and are the controlling hand. The Olympic Development Authority's delivery of London's Olympic Park is a case in point. The partnership and close working relationship with its delivery partner achieved an unprecedented success in bringing the construction of the project to completion with much lower levels of injury and ill health than might be predicted for a project of that size and complexity (Shiplee *et al.*, 2011; Bolt *et al.*, 2012).

Culture

The tradition of bad practices has left many professionals, those charged with the task of bringing projects to completion, ill-equipped and poorly supported to autonomously make and implement the ethical decisions necessary to ensure good practices within the industry. It is in this aspect of an industry's culture, its 'safety culture', where workers face the risk of harm that had become the focus of attention of those working in the field of occupational safety and health. A range of approaches has been adopted to manipulate that culture to achieve improved outputs in safety compliance, reduced accidents and a positive safety leadership from senior management and board teams. The works of Geller or Cooper on workplace behaviours are internationally recognised and have been adopted by some statutory safety authorities (see e.g. Fleming and Lardner, 2002; Health and Safety Authority, 2013). However, the issue is problematic, not least because there is no universally accepted definition of what constitutes a 'safety culture', much less what it should look like (McSweeney, 2002; Tijhuis and Fellows, 2012; Hofstede, 2014).

A common phrase used to explain safety culture is 'it is the way we do things here', but this is simplistically inadequate in that it describes no more than the subjective actions of people in the workplace without consideration of the contexts in which work occurs and the motivations of people to perform in that manner. The centrality of organisational culture to OSH remains contemporary (Zwetsloot and Steiger, 2015), yet these perspectives are fundamentally a behaviourist approach without an appreciation of the character of the individual workers and their system of motivations, which underlies but is not identical with their behaviour (Fromm, 2003). Schein (1995) describes culture as emerging from the beliefs, values and norms of the founders and leaders of a company and, in the case of the client, such influence on the behaviour of contractors is determinative. However, this perspective is insufficient to explain that the culture of a project and the beliefs and values of the client are not identical with those of the contractors or employees.

Cultural theories such as those espoused by Geertz (1973), Freire (1973, 2013) and Gajendran *et al.* (2012) permit a different perspective which allows the interconnectivity of Man and culture. Culture is not just about behaviours; Freire has described culture as Man's transformation of nature by his work. This transformation is the outworking of a conscious relationship with the world in which Man *'organises himself, chooses the best response, tests himself, acts and changes himself in the act of responding'* (Freire, 1973: 3). He is both in and of the world. In this there is a dialectical relationship in which Man is both the embodiment and maker of culture, a duality that negates definitions of culture as abstract matrices in which Man merely moves. There is a direct relationship between Man as culture maker and Man as builder of his or her own environment. The function of construction is to build sustainable environments that satisfy the needs of those who carry out the construction and those who use and maintain the finished project. As a system of uniquely human controls culture and morality are identical and an organisation's culture is the dynamic outworking of reasoning and the perception of agency by stakeholders.

Morality and ethics reasoning

Geertz's (1973) description of culture means that morality and ethics reasoning informs a rational understanding of what Man should or should not do. This is fundamental to culture and presupposes that Man is a self-reflective agent capable of decision-making and with the capacity to act independently. Morality and ethics is concerned with guiding behaviour such that at a fundamental level that behaviour is non-injurious to others and in its more evolved forms it actively contributes to the good of others. For example, the triple objectives of the sustainability of any business are for its activities to be good for the individual worker, good for the company and good for society. In this we have echoes of Fromm (2003), who contends that in order to determine what is fundamentally good for the individual we must first determine what is fundamentally good for humanity, which necessarily is good for the individual. It is not the corollary that what is good for the individual, or good for the company, is necessarily good for humanity, and the failure to appreciate this distinction lies at the heart of organisational self-interest and the negation of agency. Thus the third paradigm in the Seoul Declaration (ILO *et al.*, 2008: 5) recognises the proactive ethic in its statement: '*...occupational safety and health requires a fundamental conceptual shift towards the creation of a culture enhancing workers' well-being and welfare, away from a myopic focus on responsive accident-prevention activities.*'

This distinction illustrates a differentiation between a morality that is mandatory – 'you must not harm others' – an absolute negation of autonomous agency, and one that is desirable but ultimately non-obligatory – 'you should do good' – a partial recognition but not full acceptance of autonomy. Piaget and Kohlberg (cited in Partington, 1997) explored the development of moral reasoning to understand how moral choices were arrived at. Piaget saw two distinct stages that take place within a social context: heteronomous or constrained morality and autonomous reasoning. Kohlberg advanced Piaget's work (Crain, 1985) and his stages of development of ethics reasoning commences with pre-conventional reasoning informed by fear of punishment and self-interest. Conventional moral reasoning stems from group and societal conformity. Post conventional reasoning is based, at the penultimate stage, on the recognition of universal rights and abstract moral principles and, at the ultimate level, morality is based on the recognition and acceptance of equal existence of all living beings. In these we see the transition from constrained moral choices to autonomous decision making.

Eckensberger (2007) introduced two social interpretation spheres: interpersonal defined by concrete interactions with concrete persons; and transpersonal determined by functions and roles. As a level of client maturity the latter is indicative of the moral agent reasoning beyond personal or organisational considerations towards universal principals, reflecting the needs of users, whereas the former necessitates client empathy with and reciprocal respect for the end user and is akin to the ultimate stage identified by Kohlberg (Crain, 1985). This resonates with the relationships that exist between client, contractor and society, and the levels of cultural maturity demonstrated by individuals and groups. In the contract relationship, ethics reasoning functions at the conventional level of societal conformity and is based on a compliance with the terms of the contract and adherence to the rules of society. Notwithstanding that the terms of the contract, in compliance with national and international laws, are nonetheless drafted and interpreted from the perspective of group interests, the interest of the respective members of the delivery chain have precedence over the wider interests of society, though these are considered in as far as the legal obligations demand. It becomes difficult in these circumstances for those whose reasoning is based on concepts of universal justice to demand that the parties to the contract act in a way other than adhering to legal and contractual obligations, that is, self or group interests supported by law.

Egan (1998) discussed the role of a competent and sustainable construction industry, achieved through commitment at all levels, where:

- clients and their design teams work in partnership with the contractors and enforcement agencies;
- clients commit to making sustainability work for all those affected by their work;
- clients will engage contractors who will act and will employ those who will act as they would expect others to act; and
- business strategies and objectives are prefaced with a commitment that goals will be achieved in a manner that prevents harm and contributes to the well-being of workers and end-users.

In that regard an evolved client is one that has the intellectual and technological capabilities to prevent loss or harm, is capable of meeting the needs of their contractors, employees and the society within which they operate, and can contribute to the well-being of their stakeholders without risking the sustainability of the company.

Ethics reasoning lifts the individual from the constraints of rules-based codes of ethics, providing them with the ability to reason out the ethical issues in any given scenario, including public policy issues, contractual obligations and constraints, and social duties to others, public and end users, affected by their decisions.

Praxis in occupational safety and health

The philosophy of praxis (McAleenan, C., 2016) as practiced in workplace safety and health sets a standard far beyond the quality paradigm in as much as quality relies upon a rigid consistency of approach to deliver a predetermined outcome. What sets praxis apart from quality theories is that critical reflection takes place at each point of checking, reflecting that change has already taken place and that the journey may also have changed (Figure 4.3).

The rigidity within quality models exposes their fixation with a consistency of approach and fails to contextualise with society other than as a commodity. Profit and economic viability drive the changes without reference to societal impact or the responsibility of key stakeholders. Conversely dialogics recognises society as an integral aspect of what is produced, establishing that workers and other stakeholders analyse the impact of the work on themselves, their colleagues compelling them to consider the wider harm the work and the product may have on society. In South America, where many countries are undergoing substantial economic expansion, mega-projects serve as the primary enabler of other economic activities. In Amazonia, for example, the proposed creation of seventeen large-scale and hundreds of mid-scale hydroelectric dams poses a number of threats including the industrialisation of the jungle, territorial reorganisation, loss of biodiversity, deforestation and the potential collapse of the hydrological function of the Amazon basin (Little, 2013). In Amazonia the primary clients, governments of the constituent states, and the secondary clients, major international investors, have an obligation to those affected by these projects to ensure that the economic development and subsequent construction projects take place in conjunction with respect for the rights of the people affected and the biological world, the Amazon Basin being vital to the health of the planet. For the workers harm comes in many ways and protection of their livelihood will determine the objective meaning behind decisions they make. In the Amazonia projects the efforts of workers for gaining better working and living conditions have not been linked to environmental or rights agendas as they are tied inexorably to the projects that are destroying the region. This creates a situation of powerlessness and alienation of the workers, until through praxis they grasp the opportunity to recognise and address the dilemmas of competing objectives. Through praxis contractors and clients have a tool to define, understand and appropriately address the societal consequences of construction projects.

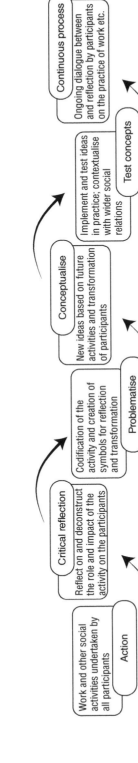

Figure 4.3 The praxis model for organisational management

Action
Work and other social activities undertaken by all participants

Critical reflection
Reflect on and deconstruct the role and impact of the activity on the participants

Problematise
Codification of the activity and creation of symbols for reflection and transformation

Conceptualise
New ideas based on future activities and transformation of participants

Test concepts
Implement and test ideas in practice; contextualise with wider social relations

Continuous process
Ongoing dialogue between and reflection by participants on the practice of work etc.

Core criteria for evaluating cultural maturity and ethics

The clients' duty to work from a solid ethical foundation stems from external obligations to the end user and the internal obligations to their delivery team. For example, delving into the intent behind the third paradigm in the Seoul Declaration (ILO *et al.*, 2008) and exploring some of the relevant definitions the following is held to be true:

- A competent company is one wherein the strategy, the managerial structures, polices and the way in which the client acts to meet its responsibilities towards all the stakeholders combine in a way that:
  - ensures the safety of workers and users;
  - enhances the quality; and
  - satisfies fiscal sustainability.
- Cultural maturity is when a client demonstrates that it has the necessary attributes essential to achieving success in health and safety, quality and environmental considerations in construction while meeting its obligations towards all of its stakeholders (OECD, 2004).

McAleenan and McAleenan (2009) developed an analytical process for determining the cultural maturity of organisations. A number of core criteria are identified that, considered in their totality, the absence of one or more will severely impair the company's sustainability in times of economic crises. This would have a negative impact on its ability to remain viable relative to competitors in times of economic stability. The criteria are:

- ethics reasoning;
- agency;
- personal and corporate competence; and
- organisational sustainability.

Criteria 1: Ethics reasoning

The success of a company, whether it is measured in economic terms or in meeting its social obligations to those who work for it, is dependent upon the company achieving cultural maturity levels akin to the optimum levels of ethics reasoning.

Notwithstanding the OECD (2004) principles of corporate governance requiring corporations to conduct their business with financial integrity and in a manner that respects their key stakeholders interests, it is an indicator of cultural and ethical maturity that businesses, no less than individuals, act morally and from optimum levels of reasoning. The fact that some businesses do not act in this fashion is not a negation of the integrity and maturity of the many individuals that comprise that business at any given point. The dominant forces in construction, as with any other industry, are the regulatory structures within which construction takes place, the financial arrangements for projects (including banking and private investment), the client for whom projects are constructed and those with ownership and shareholder rights in the various contracting companies. Their experiences and interpretations of cultural manifestations are unique as are those of employees but with the added qualifier that their decisions and actions carry a force that outweighs those of "subordinates" in the chain of authority. The worker is constrained by the decision-making authority of the employer and those appointed to manage on his or her behalf, and they in turn are constrained or influenced by the requirements of the client. Ethics reasoning requires that the organisation recognises and supports the decision-making capacity of autonomous agency within the workforce.

Criteria 2: Agency

Though required to balance self-interest with social interests, employee agency appears counterintuitive to employer and employee alike, hence the activities of workers are influenced by external and internal constraints. It does not follow that the requirements of the employer are followed unthinkingly or in agreement, thus pronouncements to work safely are interpreted individually in the context of time pressures, perceptions of employees and budgetary constraints. All of these may be influenced by factors beyond the construction site and other cultural signs in the environment, including community perceptions, the future functions of the structure and the personal histories of the individual.

The stages of moral development are universal and applicable to all individuals. By this rationale clients and their agents are capable of attaining those highest stages of universal moral principles that support the rights of all.

Criteria 3: Personal and corporate competence

The traditional thinking on competence defines it as having knowledge, skills and experience, but that is a restrictive definition which ignores some fundamental requirements, namely resources and authority to make decisions. Garavan (1997) noted that the Irish courts had also looked at the matter as far back as 1977, where the Supreme Court, commenting on the qualities of a competent person in the case of *Dalton* v. *Frendo* (1977), held that: '*[having] due regard to the age, skill and experience of a worker, he or she will know the hazards associated with their work and be able to apply the controls necessary to prevent harm.*' It was established that the ability to work safely is an integral aspect of competence and needs to be recognised as such in the execution of safe systems of working. There is widespread agreement that a competent worker is skilled, authoritative and in control of his work (ISSA, 2003; ANSI/AIHA, 2005; ILO *et al.*, 2008; Ayers and McAleenan, 2008).

In terms of practice and skill, Philip McAleenan (2016) and Ciaran McAleenan (2016) have re-evaluated competence on the basis of the many legal definitions that are incorporated into national statutes and guidance, and deconstructed it to arrive at a more appropriate definition of competence wherein the competent company respects the autonomy of the worker, team and company in decision making. The impact of this on safe working is to place control of safety directly into the hands of those responsible for carrying out the work activities. This definition meets the Quebec Protocols (ISSA, 2003) for reintegrating safety into work competence, ensuring that the worker-centric approach is properly embedded.

Criteria 4: Organisational sustainability

Organisational sustainability is concerned with the triple objectives of meeting the needs of the company, the employee and society, which are accorded equal respect and consideration. It follows that as uncertainty increases so too will the resistance to change and the justification of inequality (Graham and Estes, 2012). Inequalities manifest themselves in the ability of contractors to survive and in the early years of the recession of 2008 the larger contractors generally were better able to take the hit through cutbacks, layoffs and wage reductions. When smaller contractors, more heavily in debt to the banks, collapsed, with the loss of their plant and equipment, they were poorly placed to take advantage of any opportunities that did occur. Competition was fiercer with larger companies tendering for lower-cost contracts normally sought by smaller contractors, who may then have been engaged to conduct the work of the contract at lower fees than would have been the case had they been in a more equitable position.

Businesses, particularly large corporations that are influential on smaller companies, are structured along and succeed as a result of conservative practices (Jost et al., 2003). The organisational structure defines the roles and responsibilities of managers, defines the relative status and norms of behaviour of staff and line functionaries, mechanisms and processes for allocation of resources and in general sets the rules for the game. This suggests that those most likely to rise in the business world are those whose world view is conservative, at least in the sphere of economics. Hofstede's (2014) work on culture suggests that organisational cultures, as opposed to national cultures, are more manageable because they are based on practices rather than underpinning values. However, following the rules of the game does not imply that there is a specific organisational culture that dominates, and, whereas those who rise in the managerial hierarchy may be seen as following an occupational set of rules rather than organisational rules, they are nevertheless individuals within an organisation that hosts '*multiple, dissenting, emergent, organic, counter, plural, resisting, incomplete, [and] contradictory cultures*' (McSweeney, 2002: 96). Thus conservatism may in these cases be behavioural rather than ideological but nonetheless impacts upon the organisation's ethical foundation. A conservative world view would behave in the same old way, even when it proves to be disastrous to do so. However, Graham and Estes (2012: 41), reflecting on research conducted from the 1950s into the biological roots of ideology, concluded that: '*some of the defining aspects of conservative ideology – resistance to change and justification of inequality – were motivated by deep seated psychological needs to manage uncertainty and threat.*'

In times of uncertainty it is those qualities of the stable business and those who run them that are least suited to effecting the necessary changes that will meet the challenges of instability. The failure to recognise the need for innovative approaches has led many companies to fail during times of economic crises.

Strategies designed solely to deliver success in times of economic stability are fundamentally flawed in that when circumstances change or new conditions manifest themselves the strategies may fail. What is required is evolution of the ability to develop novel strategies, and this lies with those of a liberal ideological root, which is (ordinarily) those who are not at the helm of organisations. For example, Semco, a company facing closure during Brazil's recession in the 1980s, introduced a more liberal agenda that restructured the company, while recognising and encouraging agency (Semler, 1999). This was made possible by creativity at board level and success was achieved by the innovativeness and resourcefulness of the workforce that took account of the needs of society and the requirements for an ethical supply chain.

Conclusion

In reflecting upon the ideological basis of ethics when discussing the need for clients to work from a solid ethical foundation, the critical components – reasoning, agency, competence and organisational sustainability – are core considerations. In as much as success comes from an ethically derived and morally delivered project as it does from financial fulfilment, the safety, health and well-being of all those impacted is valued and catered for in total design and construction considerations. Cultural maturity recognises that business progression and societal responsibility are intertwined and should be demonstrably so. An organisation's culture, within which its ethical codes exist, is more than a slogan on the company vision statement. It is a complex and poorly understood concept, with the construction professions tending towards viewing it as a concrete object-in-itself capable of manipulation to achieve particular ends, for example improved performance and zero accidents. However, this tendency is not an amenable conceptualisation of culture to shed meaningful insight into the ills of the construction industry. Culture is the context in which Man exists, no less so in the workplace than in society at large. The individual, whether client, contractor or end-product user, will interpret cultural manifestations, behaviours, institutions, instructions and so on, and respond in his or her own unique fashion, including within the constraints imposed by the circumstances. Though superficially it may appear that many are responding identically, fundamentally that is not the case.

Though neither the culture nor the cultural maturity of the workplace is determined by the client, the client is ultimately responsible for the ethical foundations underpinning the projects that they commission. The establishment of a positive ethics foundation within which ethical considerations are appropriately addressed and moral behaviour delivered is a prerequisite to a sustainable preventative culture. Concomitant with this are programmes to enhance the ethics reasoning of professionals: those who will be influential on the projects, who will be cognisant of and responsible for meeting the requirements of the client and social needs of the community that live in the built environment; and a recognition of and support for agency in the delivery chain. Together these contribute to the ethical and cultural maturity of client companies.

References

Abdul-Rahman, H., Karim, S. B. A., Danuri, M. S. M., Berawi, M. A., and Yap, X. W. (2007). Does professional ethic affects construction quality? In: Yunus, Y. M., Sudin, H. A., Suhaimi, M. S. N. M., Haron, R. C., Din, A. M. and Osman, J. (Eds.). *Enhancing empowering the profession proceedings of the Quantity Surveying International Convention 2007 (QSIC 2007), 4–5 September 2007, Kuala Lumpur, Malaysia.*

Abdul-Rahman, H., Wang, C. and Yap, X. W. (2010). How professional ethics impact construction quality: Perception and evidence in a fast developing economy. *Scientific Research and Essays*, Vol. 5 (23), 3742–9.

ANSI/AIHA (2005). *American National Standards for Occupational Health and Safety Management Systems [ANSI/ AIHA Z10].* Falls Church, VA: American Industrial Hygiene Association.

Ayers, G. and McAleenan, C. (2008). Encouraging meaningful and effective consultation about occupational health and safety (OHS) in the construction industry: A recognition of workforce competence. In: *Proceedings of XVIII World Congress on Safety and Health. Safety and health at work: A societal responsibility, Seoul, South Korea. 29 June–2 July 2008.* Seoul: KOSHA.

Bangladesh RMG (2013). *Accord on fire and building safety in Bangladesh* [online]. Available at: http://bangladeshaccord.org/wp-content/uploads/2013/10/the_accord.pdf (Accessed 24 March 2017).

Bolt, H., Haslam R., Gibb, A. and Waterson, P. (2012). *Pre-conditioning for success: Analysis of human and organisational factors, ODA, Lessons learned from the London 2012 Games construction project.* London: Olympic Development Authority.

Crain, W.C., (1985). *Kohlberg's stages of moral development: Theories of development.* Upper Saddle River, NJ: Prentice-Hall.

Dorman, P. (2012). *Estimating the economic costs of occupational injuries and illnesses in developing countries: Essential information for decision-makers.* Geneva: International Labour Organization. Available at: www.ilo.org/wcmsp5/groups/public/---ed_protect/---protrav/---safework/documents/publication/wcms_207690.pdf (Accessed 24 March 2017).

Eckensberger, L.H. (2007). Morality from a cultural perspective. In: Zheng. G., K. Leung and J. G. Adair (Eds.). *Perspectives and Progress in Contemporary Cross-cultural Psychology*, 25–34. Beijing: China Light Industry Press [online]. Available at: http://iaccp.org/ebook/xian/PDFs/2_2Eckensberger.pdf (Accessed 24 March 2017).

Egan, J. (1998). *Rethinking construction: The report of the Construction Task Force.* London: HMSO.

Fleming, M. and Lardner, R. (2002). *Strategies to promote safe behaviour as part of a health and safety management system.* London: HSE Books.

Freire, P. (1973). *Education: The practice of freedom.* London: Writers and Readers Publishing Co-operative.

Freire, P. (2013 [1974]). *Education for critical consciousness*. London: Bloomsbury Publishing Plc.

Fromm, E. (2003). *Man for himself: An inquiry into the psychology of ethics*. London: Routledge Classics.

Gajendran, T., Brewer, G. Dainty, A. and Runeson, G. (2012). A conceptual approach to studying the organisational culture of construction projects, *Australasian Journal of Construction Economics and Building*, Vol. 12 (2), 1–26.

Garavan, T. N. (1997). *The Irish Health and Safety Handbook*. Dublin: Oaktree Press, pp. 69–74.

Geertz, C. (1973). Thick description: Toward an interpretive theory of culture. In: Geertz, C. (Ed., 1973). *The interpretation of cultures: Selected essays*. New York: Basic Books, pp. 3–30.

Gibson, O. (2014). More than 500 Indian workers have died in Qatar since 2012. *Guardian online*. Available at: www.theguardian.com/world/2014/feb/18/qatar-world-cup-india-migrant-worker-deaths (Accessed 24 March 2017).

Graham, J. and Estes, S. (2012). Political Instincts. *New Scientist*, 3 November 2012, 41–3.

Haugbølle, K. and Boyd, D. (2013). *Clients and users in construction: Research roadmap report*. CIB Publication 371. Rotterdam: CIB General Secretariat.

Health and Safety Authority (2013). *Behaviour Based Safety Guide*. Dublin: HSA.

Health and Safety Executive (2015). *Costs to Britain of workplace fatalities and self-reported injuries and ill health, 2013/14*. London: HSE.

Hofstede, G. (2014). *Organisational Culture* [online]. Available at: https://geert-hofstede.com/organisational-culture.html (Accessed 24 March 2017).

ILO (2012). *Estimating the economic costs of occupational injuries and illnesses in developing countries: Essential information for decision-makers*. Geneva: International Labour Organization.

ILO, ISSA and KOSHA (2008). *Seoul declaration on safety and health at work*. XVIII World Congress on Safety and Health. *Safety and health at work: A societal responsibility, Seoul, South Korea. 29 June–2 July 2008*. International Labour Organisation, International Social Security Association and Korean Occupational Safety and Health Agency, Seoul: KOSHA.

ISSA (2003). Québec City protocol for the integration of occupational health and safety (OHS) competencies into vocational and technical education. *2nd International Seminar on Occupational Health and Safety Training, October 2003, ISSA Quebec, Canada*. International Safety and Security Association – International Section on Training and Education, Quebec, Canada: International Safety and Security Association.

Jost, J., Glaser, J., Kruglanski, A., and Sullowat, F. (2003). Political conservatism as motivated social cognition. *Psychological Bulletin*, Vol. 129 (3), 339–75.

Little, P. (2013). *Mega-projects in the Amazon: A geopolitical and socio-environmental analysis with proposals for better governance for the Amazon*. [Online] Available at: www.dar.org.pe/archivos/publicacion/145_megaproyectos_ingles_final.pdf (Accessed 24 March 2017).

McAleenan, C. (2016). *Operation analysis and control: A paradigm shift in construction safety management. PhD dissertation*. Cardiff: University of South Wales.

McAleenan, C. and McAleenan, P. (2014). The application of deductive logic to determine the objective conditions impacting upon cultural maturity. In: Aulin, R. and Ek, A. (Eds. 2014). *Achieving sustainable construction health and safety proceedings of CIB Working Commission W099, Lund Sweden 2–3 June 2014*. Rotterdam: ICONDA CIB Library.

McAleenan, P. (2016). *A novel approach to health and safety in construction: culture, ethics reasoning and leadership. PhD Dissertation.* Cardiff: University of South Wales.

McAleenan, P. and McAleenan, C. (2009). Development of the competent company in the context of the Seoul Declaration: Prevention through global partnerships. In: *Proceedings of Canadian Society of Safety Engineers PDC. Calgary Canada, 20–23 September 2009.* Toronto: CSSE.

McSweeney, B. (2002). Hofstede's model of national cultural differences and their consequences: A triumph of faith – a failure of analysis. *Human Relations*, Vol. 55 (1), 89–118.

OECD (2004). *OECD Principles of Corporate Governance.* Policy Brief [Online]. Available at: www.oecd.org/corporate/oecdprinciplesofcorporategovernance.htm (Accessed 24 March 2017).

Partington, G. (1997). A critique of Piaget and Kohlberg. *International Journal of Social Education*, Vol. 11 (2), 105–19.

Qatar Foundation (2012). *QF mandatory standards of migrant workers' welfare for contractors and sub-contractors.* [Online] Available at: www.qf.org.qa/app/media/2379 (Accessed 24 March 2017).

Rolnik, R. (2009). *Report of the Special Rapporteur on adequate housing as a component of the right to an adequate standard of living, and on the right to non-discrimination in this context.* [Online] Available at: https://documents-dds-ny.un.org/doc/UNDOC/GEN/N09/446/ 64/PDF/N0944664.pdf?OpenElement (Accessed 25 March 2017).

Schein, E. H. (1995). *The leader of the future.* Drucker Foundation volume on Leadership. Cambridge, MA: MIT Sloan School of Management, 1 April. Available at: http:// dspace.mit.edu/bitstream/handle/1721.1/2582/SWP-3832-33296494.pdf (Accessed 24 March 2017).

Semler, R. (1999). *Maverick.* London: Random House Publishers.

Shiplee, H., Waterman, L., Furniss, K., Seal, R. and Jones, J. (2011). Delivering London 2012: Health and safety. In: *Proceedings of the Institution of Civil Engineers: Civil Engineering*, Vol. 164 (5), 46–54.

Tijhuis, W. and Fellows, R. (2012). *Culture in international construction.* London: Routledge.

Zwetsloot, G. and Steiger, N. (2015). *Towards an occupational safety and health culture.* [Online]. EU-OSHA, OSHWiki. Available at: https://oshwiki.eu/wiki/Towards_an_ occupational_safety_and_health_culture (Accessed 24 March 2017).

Part II

Governance

Processes and mechanisms

5 A review of funding and its implications for construction clients

Abdul-Rasheed Amidu

Introduction

Across the globe the construction industry comprises both private and public clients engaging in four major categories of construction projects: industrial, infrastructure, residential and commercial. These construction projects, by their nature, are extremely complex and dynamic human endeavours that involve the dilemma of funding them in an efficient and profitable manner. This has risk implications not only for what can be produced by way of a building or infrastructure but also on the process of undertaken projects. This chapter presents financing models for construction projects and explores their implications for clients. Although primarily focused on private projects, more and more public projects are procured using these models. The funding of public projects through the exchequer still requires a business case to be produced and a justification for the money delivered. However, the greater requirement for accountability of public funds both restricts the freedom to act efficiently and increases the administrative burden through the project in order to deliver accountability. To represent this, the chapter will explore Private Finance Initiatives (PFI) and other public–private partnerships. In particular, the chapter will consider the risk implications of the funding on the types of procurement. The chapter does have a UK focus where substantial amounts of client construction takes place under a private model. In many other countries, major construction is funded directly through the state where the delivery of the project is of greater importance than its financial efficiency. However, in these countries, as governments seek greater control, tighter financial approaches will be instigated.

Financing models for construction projects

Corporate finance versus project finance

The literature suggests a number of different categorisations of forms of financing a new construction project. One of these is the distinction between corporate and project finance. As shown in Figure 5.1, these two forms of finance refer to a company's capital structure, which consists of the equity and debt used to finance

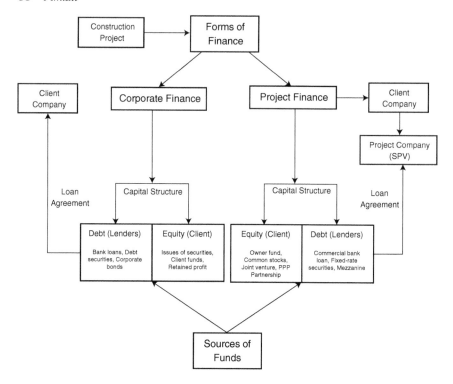

Figure 5.1 Financing models for construction projects

a project, but are fundamentally different in terms of the nature of corporate support or other kind of security sponsors provide to finance the debt component. In other words, they describe where the sponsor (or client) liability ends in case of non-payment. Project sponsors, in a construction context, are generally the project owners (or clients) with equity stake in the project. A construction project can also be sponsored by a single company or a consortium of companies.

Traditional corporate finance is a financial arrangement where sponsors of a new construction project use all of the assets and cash flows from their existing company to guarantee the additional credit provided by lenders (Gatti, 2008). Thus, in contrast to project finance, the lender does not primarily depend on the feasibility of the construction project itself or the revenue stream generated by the project to determine the potential risk of lending, but relies substantially on the sponsors' creditworthiness or the value of their company's physical assets (Merna and Njiru, 2002). From the project sponsor's perspective, corporate debt financing is generally less costly in comparison to project finance, but could have severe negative repercussions if the project is not profitable or fail completely – the lenders will have full recourse (or claim) to the sponsors company's assets (Figure 5.1).

Project finance, on the other hand, is believed to have originated in the energy generation sector (Gatti *et al.*, 2007) and was first used on a large scale to exploit the North Sea oil during the 1970s, when the scale and risk of the investment to be undertaken far exceeded the capability of any single company or a single consortium of companies (Kleimeier and Megginson, 2000). More recently, project finance schemes have been widely sought to fund complex and risky construction projects.

The term 'project finance' is defined in a variety of ways in academic literature. Writing in *Harvard Business Review*, Wynant (1980: 166) defined project finance as '*a financing of a major independent capital investment that the sponsor has segregated from its assets and general purpose obligations*'. This definition highlights one key feature of project finance: the notion that, in project finance, the sponsors of the project are removed from any obligations to repay the project debt or interest if the project fails (Buljevich and Park, 1999). In other words, the project is perceived as both legally and economically self-contained in any debt financing agreement. As argued by Brealey and Myers (2000), project finance involves financing the development and construction of a new project in which the loan agreement is linked closely to the prospect of the project, thereby minimising the risk exposure of the project sponsors. Nevitt and Fabozzi (2000: 1), on the other hand, define project finance as:

> a financing of a particular economic unit in which a lender is satisfied to look initially to the cash flow and earnings of that economic unit as the source of funds from which a loan will be repaid and to the assets of the economic unit as collateral for the loan.

In contrast with Wynant's definition, this implies that lenders not only have recourse to the project's cash flows and assets but also to the sponsors' company assets.

Special Purpose Vehicle (SPV) as mechanism

Despite the multiplicity of definitions of project finance in academic literature, the majority of authors agree that project finance is usually achieved through a specific project company. Esty and Sesia (2010) maintain that project financing involves the creation of a legally independent project company for the purpose of financing investment in a single purpose capital asset, usually with a limited life. Thus, the project company, also known as a Special Purpose Vehicle (SPV), is a new legal entity created with the sole mission of executing a specific project and the duration of an SPV is expected to equal the lifetime of the project. Established and controlled by the project sponsors, the SPV is the centre of the project through which contractual arrangements with lenders, contractors, operators and suppliers are negotiated (Vinter, 1998). Figure 5.2 provides an illustration of a typical SPV set up, which is essentially a nexus of supply and sales contracts that exist between the SPV and the various participants involved in delivering the project (Romih, 2008).

As shown in Figure 5.2, it is normally the project sponsor (s) of the SPV, who is either the main investor or a joint venture of several organisations bounded

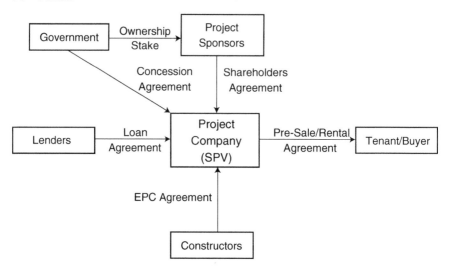

Figure 5.2 Typical set up of a private sector SPV

together under a shareholders' agreement, with strategic objectives to complete the project, extract the profit, share in the risk of carrying out the project and retain control of the project for as long as possible (Vinter, 2005). The sponsors usually invest the majority of equity in the SPV and are consequently the owners of the project. In a construction projects context, for instance, they may include construction companies, FM companies, institutional investors, investment companies, insurers and pension funds. In a large-scale infrastructure project, the government (or the procuring authority) may also wish to retain some ownership right in the project and therefore would also be a sponsor of the project. Apart from ownership, the government may also be involved in the legislative and planning process through the granting of concession rights for the construction, maintenance and operation of the project to the SPV (De Lemos *et al.*, 2000).

As noted previously, both corporate and project finance permit the use of a leveraged capital structure for the financing of the project. The level of leverage varies from project to project but a debt-to-equity ratio of 70 per cent and above is generally preferred in project finance (Buljevich and Park, 1999). As Figure 5.2 illustrates, to secure the debt financing required for undertaking a new construction project the SPV will have to enter into loan agreements with lenders who consider cash flows from the project as being the primary source of loan reimbursement and the SPV's assets as collateral. In other words, the SPV is the borrower for the project. The creation of the SPV and its role as a borrower implies that lenders have no recourse (or claim) to the sponsor's company assets. This means that, with project finance, the new project and the project sponsor live two separate lives such that the latter can shift the debt burden, operating risk and accounting liabilities to a third party, while at the same time retaining some of the benefits of the project (Gatti, 2008). However, this does not have to be the case. It is also possible for the project sponsors to borrow independently thus giving lenders limited recourse to the assets of a parent company.

Because project finance uses the SPV as a vehicle for raising non-recourse debt finance, lenders carefully assess all the possible sources of income to the project as a precondition to lending. Typically, the source of income for a construction project is the rentals or sale price for the project. The amount of the rentals or sale price is generally detailed in the pre-rental/sale agreement entered into with potential tenant/buyer of the project output. This agreement provides some degree of certainty in relation to the ability of the project to generate enough cash flow to service its debt and other operating costs and, therefore, constitutes the basis for the entire project financial arrangement (Buljevich and Park, 1999).

Once financial arrangement has been secured, the SPV then enters into an engineering, procurement and construction (EPC) agreement with one or more constructors to deliver the substantive performance of the SPV, that is, to design, construct and commission the project at a predetermined contract sum, by a certain date and in accordance with certain specifications and standards. In some cases, the sponsors of the SPV may require more participation of the constructors by way of debt contribution to the SPV. The underlying rationale for this is that, if the constructors have debt stake in the project, they will be more committed to the project and therefore deliver a higher quality project output than if they are only contracted to construct the project (Buljevich and Park, 1999).

In addition to the core project participants identified in Figure 5.2, there are a host of other professionals who are either directly or indirectly involved with the SPV. This may include, for example, the legal specialist, who is responsible for designing the contracts essential for allocating risk and responsibilities among the various stakeholders involved in the project, and the operators, who may be required to maintain the project's asset and operate it to its maximum efficiency.

In comparison with corporate finance, project finance has the advantage of insulating the financial health of the enterprise from any possible failure of the project (Vasilescu *et al.*, 2009); in this way, sponsors are able to fund the project 'off-balance sheet'. This means that project assets and liabilities do not appear on the project sponsor's balance sheet (Borgonovo *et al.*, 2010; Jechoutek and Lamech, 1995). Brealey *et al.* (1996) also noted that one of the key advantages of project finance is that it allows the allocation of specific project risks (e.g. completion and operating risk, and revenue and price risk) to those parties best able to manage them.

One major drawback of project finance is that it is more costly to structure and organise than traditional corporate finance. Klein *et al.* (1996) found that on the aggregate level, project finance is often more expensive than corporate finance due to the fact that in corporate finance creditors cannot rely on the cross-collateral cash flows and assets as is usually the case with corporate finance. Gatti (2008) further argued that project finance usually involves a great deal of time and cost to evaluate the project and negotiate the terms of contract to be included in the documentation as well as monitoring the progress of the project.

Table 5.1 Key differences between project finance and corporate finance

Factor	Project Finance	Corporate Finance
Security for financing	Depends solely on the project cash flow	Depends on the creditworthiness of the borrower
Primary source of repayment	A lender's only recourse is to the assets of the project company	A lender may have recourse to the assets of the company
Risk distribution	Spread between parties involved, although the lender primarily bears the risk of insolvency	Fully guaranteed by project sponsors and SPV
Accounting treatment	Off balance sheet (except in limited recourse project finance, where exclusive annotation is made in the financial statements)	On balance sheet, meaning that project assets and liabilities appear on sponsors' balance sheet

As approaches to debt finance, corporate and project financing share a number of attributes including the use of term loan structures and the fact that both depend on the available cash flows to service the debt contracted. However, as demonstrated above, the limited recourse that lenders have to sponsors in project finance is a major motivation for corporates adopting this approach over the corporate finance to construction projects. This and other key differences between project financing and corporate financing are summarised in Table 5.1.

Sources of funds: equity and debt

Another categorisation of finance (as shown in Figure 5.1) considers the various sources of funds that can be assessed by clients to finance a construction project. These sources are broadly grouped into equity and debt finance. Equity financing is where SPV sponsors provide funds required for a project without incurring any debt. In other words, where 100 per cent equity financing is possible, project sponsors (owners) do not have to repay a specific amount of money at any given time. According to McKeon (1999), equity funds may be private or public and in the form of common stocks offer to public investors. In a construction project, the initial equity investors typically are the direct beneficiaries from the project and will include the owners of the land where the project will take place, the purchasers of the project output and the suppliers of essential services or products to the project. The equity provided by this group of investors is generally for a very long period of time, in which case equity investors will usually have to accept delay in dividends. From this perspective, Finnerty (1996) argued that investors will only make funds available where the expected return is commensurate with the risk of the project. However, once the project has entered the operating stage and has started to show some profitability and prospect of cash dividend payments, the company or project owners may increase its capital and thus obtain further funding for the project through the issue of common stock (ordinary share) to the public and other active investors (Finnerty, 1996).

Like equity, debt provides another funding source for construction projects. Normally, lenders and investors providing a project with debt will also require the project sponsors to have some equity investment in order to ensure sponsors have enough stakes to motivate them to execute the project to a successful completion (Nevitt and Fabozzi, 1995). Depending on whether a project is in the construction phase or operating phase, there are different providers of debt capital that sponsors can utilise (Finnerty, 1996). One of these is commercial banks that currently constitute the largest sources of construction debt financing through term loans. According to Nevitt and Fabozzi (1995), a term loan is a type of business loan with more than a one-year maturity period and repayable according to a specified schedule. For project sponsors (owners), the appeal of a term loan is that it is drawn in full and, therefore, provides equity investors with a high degree of financial leverage, the key advantages of which are lower initial equity injection requirement and enhanced owner equity returns. On the other hand, committing funds for long maturity implies that lenders are naturally exposed to political risks (Sorge, 2004). Syndicates of international commercial banks are another source of bank loans for financing construction project activity. Loan syndication is a mechanism where many banks come together and each contribute a portion of the loan requested to fund a project and, whilst these are capable of raising a large amount of debt, they usually provide floating-rate debt at a higher cost that fixed bonds (McKeon, 1999).

In addition to bank loans, the fixed-rate debt market offers alternative sources of debt capital for construction projects financing. This market consists of investors who are usually willing to commit funds for medium- to long-term securities such as bonds. Fixed-rate debt instruments are traditionally issued by a company or government. Lately, the potentials of the bond market as a source of project finance has also been explored. In general, individual and institutional investors such as banks, insurance companies, pension funds, investment trusts and government agencies would only invest in this kind of capital if the issuer's financial condition and profit potential are rated (Finnerty, 1996). Rating of bonds for project financing is a new and difficult area and may, therefore, limit the capacity of raising funds from the bond market to finance construction projects. To address this difficulty, the financial enhancement rating that assesses the ability of a bond issuer to make payment is now being used to rate project bonds (Smith and Chew, 2001). Consequently, it is expected that in future the bond market will constitute one of the largest sources of funds for construction project finance.

Leasing is another source of medium- to long-term financing in which payment obligations are created between the lessor (the financing party) and the lessee (the beneficiary) of a certain project asset. Leasing is common in both developing and developed countries and it is estimated that approximately one-third of private investment in OECD countries is financed through this financing option (Finnerty, 1996). For the purpose of financing, leases may be broadly categorised into two types: financial leases and operating leases (Buljevich and Park, 1999). Financial leases are often regarded as an alternative to debt financing and referred to as a lease agreement where the lessee is obliged to pay insurance and property tax and perform routine upkeep tasks (such as maintenance and repairs) necessary to protect the leased property. In an operating lease, the lessor is responsible for performing these upkeep tasks in addition to keeping the title of the property.

There are several other sources of finance that share the attributes of both equity and debt financing, which are therefore referred to as mezzanine financing (Pratt and Crowe, 1995). These may include 'stretched' senior debt, loan and redemption premium, deep discount bonds, subordinated loans and warrants, convertible loans and convertible preference shares.

In summary, the model presented in this section not only highlights the form and sources of finance but also emphasises the nature of recourse that lenders may have to construction project clients ranging from full recourse to non-recourse. The section also highlights a key requirement of project finance, which is the creation of the SPV, through which various stakeholders (including clients) are connected to share the risk involved in carrying out a project. However, given that these stakeholders may have diverse interests, the manner in which the risk is allocated could lead to a risk of tension arising with implications, not least on the project assets, but also on the process of undertaking projects (this is further discussed in the later section on risk implications). Also, it is important to emphasise that regardless of the ultimate structure of financing adopted, most large-scale construction projects usually require the involvement of government at least in the legislative and planning process. This implies that, in construction project finance, the nature of Public–Private Partnership (PPP) contractual arrangement could have further risk implications for the project assets and process of undertaking the projects. Hence the next section examines the use of PPP in construction project financing.

PPP in construction project financing

Public–Private Partnership (PPP) has become a favourite tool for procuring different types of construction projects in both developed and developing countries around the world. It has been widely reported to be used in the finance of projects in road and railway infrastructure, public transport, power generation, water supply and sewage, airport and port developments, and, to a lesser extent, in telecommunication infrastructure and solid waste management (Auriol and Picard, 2013).

Within academic literature there are many different definitions of the concept of PPP (Table 5.2). However, a key feature of PPP projects usually agreed by all is that it typically involves a range of different collaborative arrangements between the public and private sectors to deliver a construction project in a more cost effective and efficient manner than under conventional procurement (Hodge and Greve, 2007). This is premised on the assumption that PPP has the ability to integrate the private sector's culture of enterprise in the projects procurement process. In other words, it is generally expected that, with the involvement of private sector, services could be procured in a much more efficient manner than

Table 5.2 Definitions of PPP by different authors

Author(s)	Definition
Akintoye *et al.* (2003: 461)	'a combination of resources of the public and private sectors in the quest for the more efficient service provision'
Allan (1999: 6)	'a corporate venture between public and private sectors, built on expertise of each partner that best meets the clearly defined public need to the appropriate application of resource risks and rewards'
Commission on UK PPP (2001: 2)	'a risk-sharing relationship based upon an agreed aspiration between the public and private (including voluntary) sectors to bring about a desired public policy outcome'
Grimsey and Lewis (2005: 346)	'a variety of transactions where the private sector is given the right to operate for an extended period a service that is traditionally the responsibility of the public sector'
Klijn and Teisman (2003: 137)	'co-operation between public and private actors with a durable character in which actors develop mutual products and/or services and in which risk, costs, and benefits are shared'
Van Ham and Koppenjan (2001: 598)	'cooperation of some sort of durability between public and private actors in which they jointly develop products and services, and share risks, costs, and resources which are connected with these products'

would otherwise have been possible through the public sector alone. The involvement of the private sector in project financing is also significant in the sense that it allows for an optimised risk sharing and a holistic life cycle approach to realise a project which includes one-stop planning, construction, financing, operating, maintenance and liquidation by a private contractor (Daube *et al.*, 2008). Thus in a PPP both parties are deemed to be in partnership to deliver a project or service and to bear the risk involved jointly.

There are several models of PPP with each model involving the provision of a public service facility under some combination of design, build, finance, operate, maintain, lease, own and/or transfer. Figure 5.3 presents the main types of PPP. The extent of the responsibilities of both private and public sectors can vary significantly between PPP types. However, the public sector generally retains responsibility for deciding what projects are to be provided, the quality and performance standards of these projects and the corrective actions to be taken where performance falls below expectations (Smith, 2000 as cited in Akintoye *et al.*, 2003). Figure 5.3 also shows the development of a traditional project through partnering to PPP contracting where the assets are transferred permanently to the public sector at the end of the contracted project period.

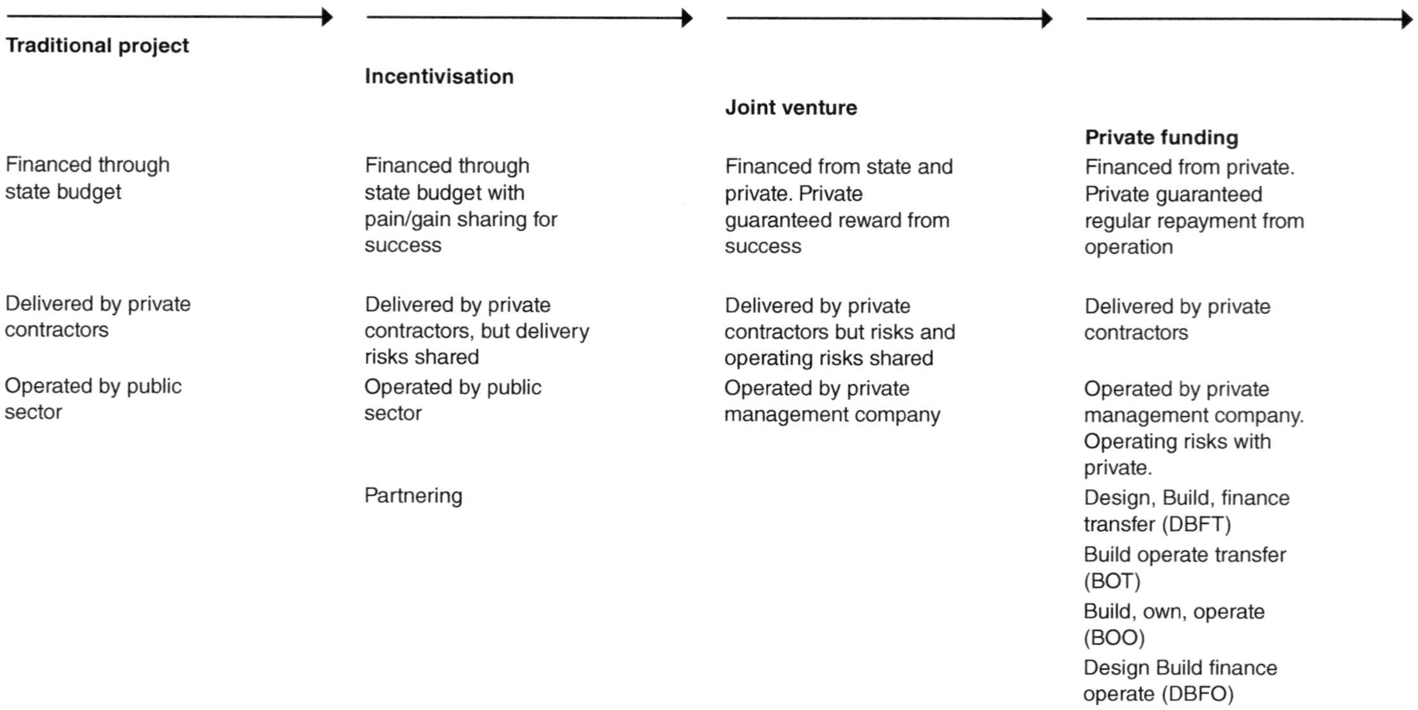

Traditional project

Financed through state budget

Delivered by private contractors

Operated by public sector

Incentivisation

Financed through state budget with pain/gain sharing for success

Delivered by private contractors, but delivery risks shared

Operated by public sector

Partnering

Joint venture

Financed from state and private. Private guaranteed reward from success

Delivered by private contractors but risks and operating risks shared

Operated by private management company

Private funding

Financed from private. Private guaranteed regular repayment from operation

Delivered by private contractors

Operated by private management company. Operating risks with private.

Design, Build, finance transfer (DBFT)

Build operate transfer (BOT)

Build, own, operate (BOO)

Design Build finance operate (DBFO)

Figure 5.3 A spectrum of funding, producing and operating public project through private means

In the UK, PPP development started in 1992 through Private Finance Initiative (PFI) – where a private sector consortium is contracted to design, build, finance and operate (DBFO) property-based services for a set period of time (typically for a minimum of 30 years). Thus, in a DBFO arrangement, the private sector is presumed to be the main provider of a 'service package' which comprises the design of any building and the accompanying operational management of the building and its aligned services, including the management of the overall financing of the entire project (Broadbent and Laughlin, 2003). In lieu of these services, the private sector supplier is assured the payment of a monthly lease cost by the public sector purchaser. Since the launch of PFI, the United Kingdom has been the undisputed leader in the use of PPP, with over 667 PPP financing projects reaching completion by 2004 (Eaton *et al.*, 2006).

From the government's perspective, one of the main advantages of raising capital through PPPs is that it helps improve focus on service delivery by reducing the time and effort that government agency spends on property- and construction-related matters (Blake, 2004). By concentrating on its core competencies, the government can also reduce upfront costs for services as they do not have to rely on their own resources in providing unfamiliar services (Cumming, 2007). The use of PPP also facilitates access to the best skills, experience and technologies from both the private and public sectors, which means that the latter is able to deliver public services more satisfactorily (Li *et al.*, 2005) while at the same time improving the quality of the services substantially (Edkins and Smyth, 2006). The private sector, on the other hand, is able to demand high profits as a precondition for investing in government infrastructure projects and consequently sharing risk with the public sector at different stages of the projects (Shen *et al.*, 2006). A further advantage is that PPP helps to reduce the lifecycle costs of projects (Li and Akintoye, 2003) and, therefore, guarantee the parties expected rate of return for their capital investment (Li *et al.*, 2005).

Although PPP offer the ability to save resources and use them more efficiently, empirical research has shown that PPP projects can sometimes run into serious difficulties due to cost overrun, unrealistic price and income projection or legal disputes among various project shareholders with differing objectives (Kumaraswamy and Zhang, 2001). Also PPP projects typically have several attributes, including the higher initial capital costs and the longer time it takes for the infrastructure to be built, which may actually make it more expensive than if the government were to borrow directly the capital needed for the project. Finally, the sponsor of a PPP project, also the client, typically faces construction and revenue risks in the use of this and other types of project financing. These issues are further explored in the next section.

Risk implications of funding types to clients of construction projects

For a particular construction project, the concept of risk may be perceived differently by different stakeholders depending on their view point, attitude and experience:

- clients (or developers) and lenders may view risk from the economic and financial perspective;
- government and related agencies from a safety and environmental perspective; and
- engineers and designers from a technological perspective (Baloi and Price, 2003).

Regardless of which perspective is adopted, risk generally represents the possibility of something happening which could have a negative impact on objectives. In a construction project, these objectives are usually stated as established targets for performance, cost, time and quality. Thus, risks are considered as failure to meet these targets.

Although most people turn to consider only the negative side of risk, academic literature actually emphasise a double-sided nature of risks both as a threat and a challenge (Flanagan and Norman, 1993) and the possibility of something occurring that may have impact on objectives. In other words, risks are not always attributable to negative outcomes – they may represent opportunities as well. This section examines mainly the negative impact of financing risks inherent in construction projects from clients' (developers') perspective. Much of these risks come from the complexity of the financial arrangement in terms of the loan structure, technical details, etc. The nature of the financial risk will also vary depending on the duration of the project. For instance, the construction phase of the project may give rise to different financing risks from those during the operating phase. Figure 5.4 illustrates the connections between different forms of finance and the way risk is distributed among the project stakeholders.

One of the main differences of the forms of finance highlighted in this chapter relates to the spread of risks between all the participants involved in the project. As Figure 5.4 illustrates, in project finance, lenders can have no (or very limited) recourse to the project sponsoring company's assets and cash flows. Non-recourse project finance is very rare in the construction industry and it is placed at the extreme end of the second finance continuum (Figure 5.4), where lenders rely solely on the creditworthiness of the project cash flows and assets. From the perspective of sponsors, this makes it possible for them to achieve higher leverage ratio than could otherwise be possible from their own corporate assets. Also, it allows sponsors to avoid any potential risk contamination; in this way, even if the project fails, the financial integrity of sponsors' core businesses would not be jeopardised as the new project will remain separate from sponsors' other activities (Esty, 2003: Sorge, 2004). A major drawback of non-recourse debt is that it exposes lenders to project-specific risks that are difficult to diversify (Sorge, 2004). In practice, therefore, lenders would usually expect some degree of corporate support in project financing. This is referred to as limited recourse project financing and the preferred financing approach to complex and risky construction projects (Vasilescu *et al.*, 2009). These three different approaches (full recourse corporate finance, non-recourse project finance and limited recourse project finance) build another finance continuum according to the level of risk

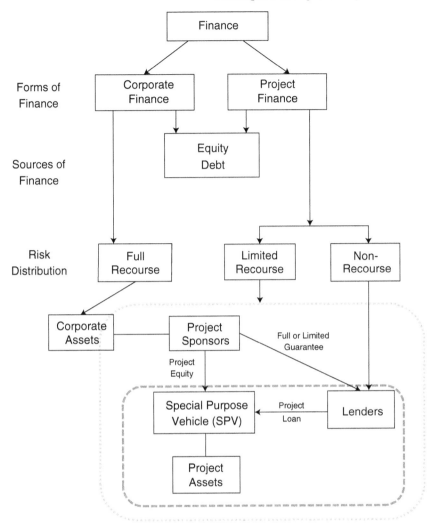

Figure 5.4 Financing models for construction projects with risk distribution

assumed by project sponsors. In other words, they describe the boundary of sponsors' responsibilities in case the project cash flows are insufficient to repay the debt contracted (Jechoutek and Lamech, 1995).

From the viewpoint of the client, there is an obvious need to ensure a robust revenue stream to support the financial arrangement for the project (Grimsey and Graham, 1997). At its simplest, the client will seek to utilise the cash flow from the project as the source of funds to pay for operating costs, cover debt financing and provide appropriate return on equity capital (Grimsey and Lewis, 2002). Thus, ensuring that the construction of the project is completed on time and to schedule is a fundamental consideration, especially to the client, who may be faced with possible financial risks arising from insufficient hedging of revenue streams and financial costs. Other construction considerations which could further expose the client to this type of risk are where the project is not completed according to standards, specifications and budgets.

This type of risk is common to any project underpinned by the creation of a highly geared SPV, and it applies with more or less force depending on whether the financial arrangement is with or without recourse to the client companies. Where there is little or no recourse to the client company, the financial risk might be low. This is because the client will be able to pass on the risk of predicted revenues not materialising to those who have provided finance of financial guarantee to the project.

Conclusion

The various forms of financing a construction project require stakeholders, including clients, to carefully consider how the financial risk of inadequate hedging of revenue streams and financing costs can be effectively managed and shared. Typically, the providers of debt finance require the cash flow of the project as the source of funds for repayment. However, financial security against the client (or sponsor) company may be sought to provide sufficient credit support. Thus, the key principle from client perspective is to achieve a financial structure where the project default risk is borne by the financiers. In this context, project finance provides an obvious choice as it enables clients to undertake projects using a highly leveraged capital structure. Using this form of finance implies that clients also need to be aware of the basic features of project finance. In particular, clients need to be aware of the fact that the conflicting objectives of the project stakeholders and the spread of risk among them may lead to the risk of tension and consequently doubts on the success of the project.

References

Akintoye, A., Hardcastle, C., Beck, M., Chinyio, E. and Asenova, D. (2003). Achieving best value in private finance initiative project procurement. *Construction Management and Economics*, Vol. 21 (5), 461–70.

Allan, J. R. (1999). *Public private partnerships: A review of the literature and practice*, Public Policy Paper No. 4. Saskatchewan: Saskatchewan Institute of Public Policy, University of Regina.

Auriol, E. and Picard, P. M. (2013). A theory of BOT concession contracts. *Journal of Economic Behavior & Organization*, Vol. 89, 187–209.

Baloi, D. and Price, A.D. (2003). Modelling global risk factors affecting construction cost performance. *International Journal of Project Management*, Vol. 21 (4), 261–9.

Blake, N. (2004). Pros and cons of public private partnerships. *The Australian Nursing Journal*, Vol. 11 (8), 15.

Borgonovo, E., Peccati, L. and Gatti, S. (2010). What drives value creation in investment projects? An application of sensitivity analysis to project finance transactions. *European Journal of Operational Research*, Vol. 205 (1), 227–36.

Brealey, R. and Myers, S. C. (2000). *Principles of corporate finance* (6th edn). Boston, MA: Irwin/McGraw-Hill Companies Inc.

Brealey, R., Cooper, I. and Habib, M. (1996). Using project finance to fund infrastructure investments. *Journal of Applied Corporate Finance*, Vol. 9 (3), 25–38.

Broadbent, J. and Laughlin, R. (2003). Public private partnerships: An introduction. *Accounting, Auditing & Accountability Journal*, Vol. 16 (3), 332–41.

Buljevich, E. C. and Park, Y. S. (1999). *Project financing and the international financial markets*. Norwel: Kluwer Academic Publishers.

Commission on UK PPP (2001). *Building better partnerships: The final report from the Commission on Public Private Partnerships, summary*. London: IPPR.

Cumming, D. (2007). Government policy towards entrepreneurial finance: Innovation investment funds. *Journal of Business Venturing*, Vol. 22 (2), 193–235.

Daube, D., Vollrath, S. and Alfen, H. W. (2008). A comparison of project finance and the forfeiting model as financing forms for PPP projects in Germany. *International Journal of Project Management*, Vol. 26 (4), 376–87.

De Lemos, T., Betts, M., Eaton, D. and De Almeida, L. T. (2000). From concessions to project finance and the private finance initiative. *The Journal of Structured Finance*, Vol. 6 (3), 19–37.

Eaton, D., Akbiyikli, R. and Dickinson, M. (2006). An evaluation of the stimulants and impediments to innovation within PFI/PPP projects. *Construction Innovation*, Vol. 6 (2), 63–7.

Edkins, A. J. and Smyth, H. J. (2006). Contractual management in PPP projects: Evaluation of legal versus relational contracting for service delivery. *Journal of Professional Issues in Engineering Education and Practice*, Vol. 132 (1), 82–93.

Esty, B. C. (2003). *The economic motivations for using project finance*. Boston, MA: Harvard Business School.

Esty, B. C. and Sesia, A. Jr. (2010). An overview of project finance & infrastructure finance: 2009 update. Cambridge, MA: Harvard Business School Teaching Note, 9-207-107.

Finnerty, J. (1996). *Project financing: Asset-based financial engineering*. New York: Wiley, Inc.

Flanagan, R. and Norman, G. (1993). *Risk management and construction*. London: Blackwell Science Ltd.

Gatti, S. (2008). *Project finance in theory and practice: Designing, structuring, and financing private and public projects*. Burlington, MA, San Diego, CA and London: Academic Press.

Gatti, S., Rigamonti, A., Saita, F. and Senati, M. (2007). Measuring value-at-risk in project finance transactions. *European Financial Management*, Vol. 13 (1), 135–58.

Grimsey, D. and Graham, R. (1997). PFI in the NHS. *Engineering construction and architectural management*, Vol. 4 (3), 215–31.

Grimsey, D. and Lewis, M. K. (2002). Evaluating the risks of public private partnerships for infrastructure projects. *International Journal of Project Management*, Vol. 20 (2), 107–18.

Grimsey, D. and Lewis, M. K. (2005). Are public private partnerships value for money? Evaluating alternative approaches and comparing academic and practitioner views. *Accounting Forum*, Vol. 29 (4), 345–78.

Hodge, G. A. and Greve, C. (2007). Public–private partnerships: an international performance review. *Public Administration Review*, Vol. 67 (3), 545–58.

Jechoutek, K. G. and Lamech, R. (1995). *Private power financing: From project finance to corporate finance*. Washington, DC: The World Bank.

Kleimeier, S. and Megginson, W. L. (2000). Are project finance loans different from other syndicated credits? *Journal of Applied Corporate Finance*, Vol. 13 (1), 75–87.

Klein, M., So, J. and Shin, B. (1996). *Transaction costs in private infrastructure projects: Are they too high?* Public Policy for the Private Sector. Viewpoint, Note No. 95. Washington, DC: The World Bank.

Klijn, E. H. and Teisman, G. R. (2003). Institutional and strategic barriers to public–private partnership: An analysis of Dutch cases. *Public Money and Management*, Vol. 23 (3), 137–46.

Kumaraswamy, M. M. and Zhang, X. Q. (2001). Governmental role in BOT-led infrastructure development. *International Journal of Project Management*, Vol. 19 (4), 195–205.

Li, B. and Akintoye, A. (2003). An overview of public–private partnership. In: Akintoye, A., Beck, M. and Hardcastle, C. (Eds.). *Public–private partnerships: Managing risks and opportunities*. Oxford: Blackwell Science Ltd.

Li, B., Akintoye, A., Edwards, P. J. and Hardcastle, C. (2005). Critical success factors for PPP/PFI projects in the UK construction industry. *Construction Management and Economics*, Vol. 23 (5), 459–71.

McKeon, P. G. (1999). High-yield debt: Broadening the scope of project finance, *The Journal of Project Finance*, Vol. 5 (3), 62–9.

Merna, T. and Njiru, C. (2002). *Financing infrastructure projects*. London: Thomas Telford.

Nevitt, P. and Fabozzi, F. (1995). *Project Financing* (6th edn). London: Euromoney Publications.

Nevitt, P. and Fabozzi, F. (2000). *Project Financing* (7th edn). London: Euromoney Publications.

Pratt, M. and Crowe, A. (1995). Mezzanine finance. *Bank of England Quarterly Bulletin*, Vol. 35, 370–74.

Romih, D. (2008). Project finance. *Journal of Local Self-Government*, Vol. 6 (2), 171–81.

Shen, L. Y., Platten, A. and Deng, X. P. (2006). Role of public private partnerships to manage risks in public sector projects in Hong Kong. *International Journal of Project Management*, Vol. 24 (7), 587–94.

Smith, A. (2000). The way forward. *The Private Finance Initiative Journal*, Vol. 4 (6), 10–12.

Smith, S. G. and Chew, W. H. (2001). Targeted risk coverage: Tempering innovation with caution. *The Journal of Structured and Project Finance*, Vol. 7 (1), 15–18.

Sorge, M. (2004). The nature of credit risk in project finance. *BIS Quarterly Review*, December, 1–12.

Van Ham, H. and Koppenjan, J. (2001). Building public–private partnerships: Assessing and managing risks in port development. *Public Management Review*, Vol 3 (4), 593–616.

Vasilescu, A.-M., Dima, A. M. and Vasilache, S. (2009). Credit analysis policies in construction project finance. *Management & Marketing*, Vol. 4 (2), 79–94.

Vinter, G. D. (1998). *Project finance: A legal guide* (2nd edn). London: Sweet and Maxwell.

Vinter, G. D. (2005). *Project finance: A legal guide* (3rd edn). London: Sweet and Maxwell.

Wynant, L. (1980). Essential elements of project financing. *Harvard Business Review*, Vol. 58 (3), 165–73.

6 Defects and insurance

Protective mechanism or driver of change?

Kim Haugbølle

Introduction

Construction is a risky endeavour with high potential losses due to errors, defects, accidents, etc. As pointed out by Haugbølle and Forman (2009), defects are one of those sticky problems that keep recurring despite various policy initiatives, development activities, etc. Defects are not insignificant problems in construction, thus a range of studies have attempted to identify the causes of defects (see e.g. Jingmond and Ågren, 2015; Sun and Meng, 2009; Love *et al.*, 2016; Hopkin *et al.*, 2016). Others have attempted to estimate the economic costs of defects. Depending on the actual definition of defects, studies estimate the economic costs as high as 5–10 per cent of the construction costs (see e.g. Josephson and Hammarlund, 1999; Love and Li, 2000; Nielsen *et al.*, 2004; Hwang *et al.*, 2009). The risks associated with defects may be dealt with in different ways:

- by contractual measures (liabilities, imposing sanctions on suppliers, etc.);
- by insurance (covers, etc.); and
- by the client (residual risks, development, etc.).

The ways clients deal with risks, in particular whether clients try to protect themselves by contractual measures, construction insurance or bear it themselves, is usually the result of long deep-rooted historical and legal contexts that differ from country to country and even within countries. The CIB W118 Research Roadmap (Haugbølle and Boyd, 2013, 2016) urges researchers and practitioners to reflect on the governance of construction, meaning the processes and mechanisms that guide the interaction of clients with construction professionals and others. Construction insurance is one of those fundamental mechanisms that govern construction, but remains largely ignored in the academic literature despite its significant role in modern society.

Insurance is a huge business in Europe employing some 1 million people in around 3,700 companies. In 2015, European insurance generated a premium income of EUR 1,200 billion and paid out EUR 976 billion in total benefits and claims. Insurance can broadly be divided into three main categories:

- life insurance;
- non-life insurance; and
- health insurance.

In 2015, the distribution of premiums on these three main categories was as follows: EUR 730 billion in life premiums, EUR 343 billion in non-life premiums and EUR 124 billion in health premiums. Non-life insurance includes motor, property, general liability and accident, with motor and property insurances providing the largest premiums (Insurance Europe, 2016a, 2016b, 2016c).

The objective of this chapter is to analyse how construction is governed by construction insurance and the implications for clients with regard to innovation and protection of long-term financial interests. Based on a constructivist perspective this chapter analyses how clients are being shaped by insurance with regard to guarantee policies after handover of a built facility – known as inherent defects insurance (IDI). More specifically, the chapter identifies construction insurance as a deep-rooted mechanism guiding:

- the interaction between clients, insurers and the industry;
- identifies two different technological frames (labelled 'protection' versus 'driver of change'); and
- explores the implications of these frames for clients.

The chapter begins with a presentation of the methodology applied in the study. It then moves on to give a brief overview of the typical approach to construction insurance across the 28 member states of the European Union. Next, it analyses the Danish Building Defects Fund as an exemplar case of an alternative insurance scheme operating as a protective mechanism as well as a promoter of change in construction. Contrasting these two frames on construction insurance, the chapter points at possible implications for construction clients.

Methodology

Theoretical framework

Understanding the role of insurance in governing construction requires a theoretical perspective that is equally sensitive to actors, technologies, processes and outcomes. This study considers insurance as a service and, as such, as an intangible technology which materially intervenes between actors of construction.

This chapter is based on a constructivist perspective emphasising that agency is governed by the intimate linkages between technical objects and social relations (see e.g. Bijker *et al.*, 1987; Bijker and Law, 1992; Oudshoorn and Pinch, 2003). Bijker (1995) introduces 'technological frame' as a theoretical concept in the analysis of interactions involving the technology within and between different relevant social groups. To describe a technological frame, Bijker proposes a tentative list of elements including goals, key problems, problem-solving

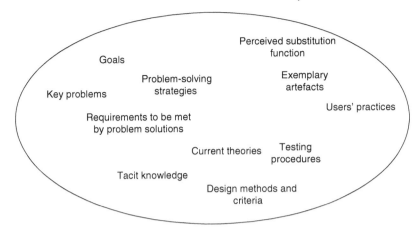

Figure 6.1 Technological frame as concept

strategies, requirements to be met by problem solutions, current theories, tacit knowledge, testing procedures, design methods and criteria, users' practice, perceived substitution function and exemplary artefacts (Figure 6.1).

Technological frame is a heuristic concept to be adapted to the actual analysis. Hence, the individual elements need to be interpreted in relation to the actual context. This implies that an element such as 'testing procedure' may be understood differently in an engineering context with regard to, for example, load bearing of structures compared to a business context with regard to, for example, checking and controlling insurance policies. This also implies that some elements may be irrelevant for some social groups, and other elements may need to be added for other social groups. Bijker (1995: 143) suggests that, since actors generally belong to more than one relevant social group, they are involved in different technological frames at the same time with different degrees of inclusion:

> The degree of inclusion of an actor in a technological frame indicates to what extent the actor's interactions are structured by that technological frame. If an actor has a high degree of inclusion, this means that she thinks, acts and interacts to a large extent in terms of that technological frame. It is expected that actors who are contemporaneously member of different relevant social groups will have different degrees of inclusion in the associated technological frames.

Based on the concepts of technological frame and inclusion, Bijker (1995: 277) suggests that sociotechnical change may follow one of three alternative configurations:

- No clearly dominant technological frame guides interactions, which tends to lead to many different radical innovations, if the necessary resources are available to the actors.

- One dominant technological frame guides interactions, which tends to lead to incremental innovations, because the stakeholders associated with the dominant frame can insist on defining both problems and solutions.
- Two or more dominant technological frames guide interactions, which tend to lead to innovations that are conventional in a double sense due to the amalgamation of vested interests.

The impact of insurance can then be analysed through technological frames in order to determine the effect that insurance has on clients and construction.

Methods and data

This chapter provides a meta-analysis of data collected from three Danish, Nordic/Baltic and European studies over the past 10 years along with ongoing interaction with the Danish Building Defects Fund over a period of more than 15 years. The Danish research project, entitled 'Defects in the Danish construction industry – strategies, action and learning', was conducted in 2006–9. The Nordic/Baltic research project, 'CREDIT – Construction and Real Estate: Developing Indicators for Transparency', was conducted in 2007–10. The European research and development project, ELIOS2 'European Liability and Insurance Organisation Schemes', was conducted in 2011–14. The author has also gained insights into the operation of the Danish Building Defects Fund from acting as a board member of one of the project developments covered by the fund.

Construction insurance as protective mechanism

This section describes the typical approach to construction insurance, in particular IDI. Hence, it is divided into three parts addressing in turn the key problems and solutions, the insurance underwriting procedure and the results in terms of IDI market developments. The use of bold in the text refers to the different elements of the technological frame. Table 6.1 provides an overview of the analysis of construction insurance as a protective mechanism.

Key problems and solutions: different types of insurance

The core **goal** of insurance is to provide financial protection in case of liabilities, damages or losses by way of transferring risks between an insured and an insurer. As stated by the European federation of insurers and reinsurers:

> Insurance protects people and businesses against the risk of unforeseeable events. It is a risk transfer mechanism by which the losses of the few are paid for by the many, with the premiums based on the risk of each individual or entity.
>
> (Insurance Europe, 2012: 5)

Table 6.1 Construction insurance as protective mechanism

	Protection frame
Goals	Transfer of risk
Key problems	Profitable insurance schemes
	Pooling risks
Problem-solving strategies	Various market-based types of insurance dependent on local legal context: TPL/PI, CAR, IDI
Requirements to be met by problem solution	Calculating premium based on type of risks, construction lifecycle, and types of constructed asset
Current theories	Market economy
Design methods and criteria	Underwriting process to set premium, terms, etc.
Testing procedures	Risk assessment procedures
	Technical inspection services
Clients' and users' practices	Heterogeneous practices: all types of clients/owners
Perceived substitution function	Compensation
Exemplary artefacts	Policy terms and conditions
	Proprietary data

The **key problem** faced by insurers is how to accurately assess the risks involved given the cover provided in order to calculate a so-called 'fair' premium to be paid by the insured to make it economically profitable to the insurer. As stated by Insurance Europe (2012: 11–12):

> On the one hand, the insurer must be able to charge a premium that is high enough to cover future claims on its pool of risks and its expenses while still making a profit. On the other hand, the amount charged to insure an individual or entity must be a sum that the insured is willing to pay and must be substantially below that of the covered amount or it would not make sense to purchase the cover.

What counts as fair pricing is clearly disputed as evidenced by the questionnaire survey by Ndekugri *et al.* (2013) with regard to all-risk insurance in infrastructure.

The **problem-solving strategy** is to offer a range of market-based types of insurances dependent on the local legal context. Construction insurance typically covers three main types of insurance (Sudres, 2011; Bunni, 2011):

- general liability policies, which are issued for each insured. These cover third party liabilities (TPL) and professional indemnity (PI);
- work in progress policies, which are issued for each construction site. These are commonly known as constructor's all risk insurance (CAR); and
- guarantee policies, after acceptance/handover. These include inherent defects insurance (IDI).

Guarantee policies after acceptance/handover may cover all types of constructed assets, but these policies are not uniformly applied. Instead, they are dependent on the type of ownership and the legal system in question. The typical policies issued for constructed assets include (Sudres, 2011):

- fire and explosion policies for each construction work (both new and existing) taken out by the owner and paid on annual basis;
- machinery breakdown policies for each construction work with regard to industrial facilities;
- owner's operations liability policy for each construction work or insured; and
- constructors' long-tail liability policies covering the constructor's liability arising from partial or total collapse due to human error and consequential damage.

This last type, long-tail liability policies, are known as inherent defects insurance (IDI). IDI is a long-term insurance covering damages to the constructed asset which result from an inherent defect discovered after completion and after the owner has taken over the constructed asset. An inherent defect is any defect in the structural works which is attributable to a defect in design, materials or workmanship. Many countries do not apply legal guarantees, and these policies are not available on an individual basis. Hence, the characteristics and implementation of this type of insurance policy is highly dependent on the local legal context. It is characterised by a single premium, paid by the insured at the beginning of the guarantee period, which covers all liability risks for the defined period, typically two, five or ten years. The insurance may also be taken out by the owner (Sudres, 2011).

With regard to the **requirements to be met by the problem solution**, Bunni (2011: 45) identifies a set of criteria for classifying risks:

- geography;
- size of project;
- legal concepts;
- effect of the risk eventuating;
- chronological origin of risk; and
- insurability.

According to Sudres (2011) the type of constructed assets is also important as differences may exist due to, for example, the types of risk and specific insurance requirements. The types of constructed asset will usually be divided into public works and civil engineering, industrial works, housing and miscellaneous works.

Insurance procedure: underwriting and use of technical inspection services

The **current theories** underlying insurance are based on neoliberal economic theory stressing the importance of competition and free markets. As highlighted by Insurance Europe (2012: 15): '*Without a competitive and innovative insurance* industry, many aspects of our modern society and economy would cease to exist or would function much less effectively.' To the extent that public regulation is required it should be 'effective' in the sense that it *'fully take into account the unique characteristics of insurance'* (Insurance Europe, 2012: 19), which, among other things, entail balanced requirements on holding sufficient capital, recognising the long-term value of insurance, freedom to differentiate between customers and freedom to insure what is insurable.

The insurance underwriting process is the crucial **method of designing and issuing** construction insurance policies which also covers various **testing procedures**. Bunni (2011: 207) identifies three layers in this process:

- The brokers, agents or sales department of an insurance company which handle and place risks within an insurance package.
- The insurer who accepts and underwrites the contract of insurance.
- The reinsurer who accepts and underwrites a number of risks accepted by insurers.

Roussel *et al.* (2015) describes in detail the underwriting process following these five typical steps:

1   Request for insurance.
2   Global check.
3   Insurer's level of interest.
4   Detailed risk assessment.
5   Terms and conditions.

Step 1 is the request for insurance by a client or owner of a built asset who may directly contact an insurer or ask an insurance broker to help obtain the best insurance conditions. The broker will typically be mandated to adapt the insurance request to the client's needs, collect and present the technical information to the insurers, and ultimately compare the offers made (Roussel *et al.*, 2015).

Step 2 is essentially a **testing procedure** as described in detail by Roussel *et al.* (2015). First, the insurer will verify that the request complies with the general scope of the insurer's business. As construction insurance requires specific underwriting competences and possibly includes a long-tail financial exposure, the insurer will initially verify if the type of cover fits into his or her portfolio. Next the insurer will check for the type of construction. Further, the underwriter will verify that the insurance cover fits with the financial capacity of the insurer. Consequently, the request may either be rejected or forwarded for a more detailed analysis internally (Roussel *et al.*, 2015).

Step 3 is another **testing procedure**, where the insurer appraises his financial interest in providing such cover based on internal policies and the financial context. According to Roussel *et al.* (2015), the selection criteria are closely linked to the insurer's risk portfolio including:

- administrative costs, including the expertise needed to study and cover a particularly market or risk;
- risk appetite, depending on the type of cover and type of construction work (specific profitability criteria based on loss experience), and depending on market foresight and financial conditions of the insurer; and
- solvency requirements, e.g. by EU regulation.

If the client passes this acceptance threshold, the insurer will move on to step 4 doing a detailed risk assessment, which forms yet another **testing procedure**. Ranasinghe (1998) points out that the risk appraisal and management procedures of insurance companies with regard to construction insurance are based on minimum acceptance criteria. According to Roussel *et al.* (2015), the insurer will appraise qualitatively the technical risks for the specific project based on his technical knowledge and determine the level of exposure. With regard to single covers specifically linked to construction works, the risk assessment made by the insurer will deal with a number of risk assessment criteria, such as materials, design, etc. (Roussel *et al.*, 2015).

Step 5 is defining the terms and conditions to be applied in the insurance policy, which can be considered an **exemplary artefact** mediating the relation between insurer and insured. Based on the risk assessment, the insurer will adapt his or her offer's terms (e.g. by excluding some risks) and conditions (e.g. by limiting some covers) taking into account the reinsurance cost. In accordance with public regulation, insurance companies are eligible to use any non-discriminatory technical criteria (Roussel *et al.*, 2015). An additional **testing procedure** is performed by technical inspection services that play an important role in relation to construction insurance and compliance with building regulation. Technical inspection services may take on two different roles: conformity assessment or risk analysis. The intervention of technical inspection services is highly dependent on the local context (Pedro *et al.*, 2010; Roussel *et al.*, 2015).

The **clients' practices** cover a diverse set of practices, as insurance is provided for a wide range of owners and construction clients holding a wide range of constructed assets that are used for different purposes. After the commencement of the insurance policy, the insured owner of a constructed asset may eventually be involved in filing a claim in accordance with the terms and conditions set in the policy. In case of such a claim, the compensation provided by the insurer represents the **perceived substitution function** for the insured owner of a constructed asset. Whether the owner considers this compensation as satisfactory substitution will depend on both the expectations of the owner and the actual cover of the policy.

Results: IDI cover in Europe

This section presents the results of the two consecutive ELIOS studies (CEA and CSTB, 2010a, 2010b; Roussel *et al.*, 2015) providing an overview of national liability and insurance systems in the European member states with regard to IDI. These can be broadly divided into two main categories:

- countries where inherent defect insurance (IDI) long-term cover has spread widely or has become mandatory (marked with black in Figure 6.2); and
- countries with no or very limited covers of long-tail liabilities or post completion (marked with grey in Figure 6.2).

According to Roussel *et al.* (2015), the presence of IDI in a country is not necessarily linked directly to the national legal schemes. In some countries, for example Denmark, Finland, France, Italy, Latvia and Spain, the existence is linked to compulsory legal requirements. In other countries, such as Ireland, United Kingdom, the Netherlands and Sweden, the existence of IDI is voluntary and market-based.

Assessing the size of the market for inherent defects insurance is difficult. Historically, statistics on construction insurance related to TPL and PI have been mixed with other types of third party liabilities (automobiles, products, etc.). Further, construction insurance is very often even mixed with property insurance.

Figure 6.2 IDI cover in European member states

According to Roussel *et al.* (2015: 109), the most reliable figures on European IDI market show that:

- France maintains a leading position with a level of direct premiums of EUR 2,500 million per year;
- the UK Home Warranty market is the second largest IDI market, with premiums around EUR 200 million annually;
- the IDI Spanish direct premiums have dropped dramatically due to the financial crisis, from EUR 364 million in 2007 to EUR 25 million in 2014;
- the Italian IDI market is generating an insurance premium of around EUR 20 million annually, even though the guarantee is compulsory; and
- the Scandinavian markets are well developed with regard to IDI, but the markets are not in the same order of magnitude as the other markets in absolute terms due to their relatively small size.

The Danish Building Defects Fund as driver of change

This section describes the establishment of the Danish Building Defects Fund as an alternative approach to liabilities and insurance. Hence, this section is divided into three parts addressing in turn the key problems and solutions, the insurance procedure and the results in terms of a significantly reduction of defects in Danish social housing. The historical account in this section rests on Nielsen (2012), the previous president of Danish Social Housing – the national association of social housing associations. Table 6.2 provides an overview of the analysis of the building defects fund as a driver of change. The use of bold in the following text refers to the different elements of the technological frame.

Table 6.2 Construction insurance as driver of change

	Driver of change frame
Goals	Improve quality of construction
Key problems	Reduce defects
	Improve maintenance
	Financial model for repairs
Problem-solving strategies	Quality assurance and liability reform of legal system
Requirements to be met by problem solution	Self-financing scheme for social housing only
Current theories	Negotiated economy
Design methods and criteria	Fixed premium of 1 per cent
Testing procedures	Third-party independent audits/compliance checks after 1 and 5 years
Clients' and users' practices	Professional and repetitive semi-public clients
Perceived substitution function	Abolish bad construction practices
Exemplary artefacts	Open access database
	Guide to better quality
	BYG-ERFA letters

Key problems and solutions: pervasive quality assurance and liability reform

Improving the quality of construction was the **goal** that became the point of departure leading to the establishment of the Danish Building Defects Fund. During the 1970s a number of social housing estates constructed as part of the great post Second World War building boom turned out to have severe problems, such as leaking flat roof constructions nurturing mould and fungi attacks. Strong public protests from the occupants eventually led to political intervention which at first included the establishment of a new public support scheme for those social housing estates being affected by building damages (Nielsen, 2012).

Within a short number of years, however, it became apparent that the problem continued to be a political menace despite economic support to remedy building damages to some 15,000 apartments in the period 1978–84. Hence, new political initiatives were taken to solve the **key problems**: remedy building damages, improve maintenance and create a new model for financing the repair of damages in the future (Nielsen, 2012).

The key **problem-solving strategy** was to launch a major quality assurance and liability reform of the legal system governing Danish construction. The reform stimulated major organisational and technological changes, which have profoundly affected construction practices and the ways in which the process of delivering construction was organised. The reform included a number of main changes (Bonke and Levring, 1996; Nielsen, 2012):

* introducing new design and performance procedures, in particular procedures of formal quality reporting during the process of design and execution, which helped to integrate the quality reporting into the construction process;
* unifying liability limitation periods in order to simplify the juridical procedures and to reduce litigation costs by introducing a common liability period of five years for all parties involved, hence replacing inhomogeneous periods varying from one year for manufacturers, over five years for consultants, to twenty years for contractors;
* establishing the Danish Building Defects Fund for social housing associations and tenants-owned cooperatives;
* introducing building maintenance guidelines; and
* introducing compulsory building inspections one year and five years after completion.

The Danish Building Defects Fund has three main tasks (Byggeskadefonden, 2016a):

* to manage and to defray the third-party independent inspections of the housing projects covered by the fund;
* to cover up to 95 per cent of the building owner's costs of repairing the damage, which have its cause in the construction of the building, if the building owner cannot make a claim against the parties involved. The fund

provides coverage for claims reported within twenty years after hand-over; and

- to disseminate lessons learned broadly in order to reduce building damages and improve quality and efficiency in construction.

A key **requirement to be met by the problem solution** was to ensure that the new building defects fund would be cost-effective and self-financing in order to avoid further public support to be spent on rectifying defects. Hence, the fund was to be financed by a premium of 1 per cent of the total value of each new construction project. Another requirement was to include social housing and tenants-owned cooperatives only. From 2000 onwards, the tenant-owned cooperatives were gradually being left out due to changes in financing schemes. Another major recent change took place in 2011 when the coverage of the fund was expanded to include larger renovations of existing social housing estates supported by the National Building Fund (Landsbyggefonden in Danish) for social housing (Nielsen, 2012).

Insurance procedure: the fund and use of technical inspection services

The **current theory** underpinning the establishment and operation of the defects fund follows closely in the wake of the Scandinavian welfare models based on Keynesian-inspired interventions to rectify market defects or limitations. In a more advanced understanding, the societal dynamics may be characterised as a negotiated economy (Pedersen *et al.*, 1992). Hence, the establishment of an organisation such as the defects fund seems to be the embodiment of what Pedersen *et al.* (1992) has labelled 'privatisation of politics', where public political responsibilities are increasingly shared with or transferred to private organisations. Hence, the building defects fund is a private organisation but essentially exercising a politically mandated public task.

With the establishment of the Danish Building Defects Fund a new **method of designing and issuing** inherent defects insurance policies was put in place, which also included various new **testing procedures**. The construction insurance procedure was quite different for the building defects fund compared to the usual underwriting procedure. First of all, the scheme was mandatory. Hence, terms and conditions were the same for all social housing projects that had to register at the fund and pay the same premium of 1 per cent of the construction costs. Second, no detailed risk assessment was conducted prior to the building project by the fund. In recent years, however, consultants have been required to file a risk assessment protocol as part of their consultancy during the works. Third, two independent third-party inspections or compliance checks were carried out one year and five years after handover based on an assessment in a five-point scale of the seriousness of a deficiency. The building owner can also file a claim with the fund. If possible, the fund will make liability claims to the responsible contractor, consultants and suppliers. The fund covers up to 95 per cent of expenditures for damage repairs that are claimed at the latest twenty years after handover (Byggeskadefonden, 2005, 2014a, 2014b, 2014c).

The building defects fund appoints an independent technical inspector to carry out the one-year inspection (effectively the inspection process starts three months after handover). All technical inspectors are required to have experience with designing social housing projects. Approximately 140 firms, typical architects and engineers, carry out the one-year and five-year inspections nationwide. The technical inspections are executed following a standardised template for assessing compliance of technical elements with building regulations and codes of conduct. The technical inspectors complete an inspection report to the building defects fund based on the project documentation and randomised physical checks of the building (Byggeskadefonden, 2014a).

The **testing procedures** have changed significantly over the years. When the fund was established, it was decided to conduct third-party independent audits only once five years after hand-over. After some years of operation, an analysis showed that the vast majority of defects were readily visible already after one year. Hence, the fund changed its testing procedure from 1997 onwards to focus on the one-year inspection rather than the five-year inspection (Nielsen, 2012).

The **clients' practices** are strongly homogenised by the fact that they all have to adhere to the same detailed public regulation on publicly supported housing. Although social housing associations and companies may try to differentiate themselves in relation to other associations and companies, they essentially share many of the same characteristics with regard to the legal context, funding schemes, types of tenant, institutionalised tenants' democracy, etc. It also characteristic that the social housing companies repetitively enters into construction and renovation projects and can be considered highly professional clients, although technically the client is the individual housing estate rather than the housing company due to the formalised tenants' democracy constituting the social housing sector in Denmark (see e.g. Scanlon and Vestergaard, 2007).

The **perceived substitution function** is avoiding defective construction practices and incompetent builders. Hence, a distinct characteristic of the building defects fund is the extensive dissemination of lessons learned on bad construction practices. Since 2005, the fund has strongly increased its dissemination activities and devoted a considerable amount of resources to organising and participating in courses, meetings, speeches, news articles and annual technical summaries. These activities are often organised in collaboration with the trade organisations and others in the construction industry.

Three **exemplary artefacts** stand out as particularly important in this respect. First, the fund operates an open access database, with access to detailed information on defects in the projects covered by the fund (Byggeskadefonden, 2016b). Second, the fund has summarised twenty-five years of experiences on defects in the 'red book' with examples and practical guidance on how to avoid replicating defective practices (Byggeskadefonden, 2014d). Third, along with a number of knowledge institutions, the Building Defects Fund for Urban Renewal and the trade association for pensions and insurance companies, the building defects fund is also highly active as a board member of BYG-ERFA along with participation in the standing technical committee. BYG-ERFA has since 1977 gathered and distributed experiences on construction practices. The experiences are conveyed in brief experience leaflets, with buildable solutions for new construction and building renewal in order to prevent defects and to remedy damages in the most appropriate manner (BYG-ERFA, 2016).

Results: quality improvements

The establishment of the building defects fund (and a similar fund for urban renewal in 1989) has been a remarkable success. Since its establishment in 1986, the fund has insured close to 250,000 dwellings in new developments and some 100,000 dwellings in renovation projects. It is well documented that the measures implemented have helped to significantly reduce the level of serious construction defects in the monitored buildings from 30 per cent at the beginning of 1990s to a few per cent in the 2010s (Figure 6.3).

The main results of the Danish Building Defects Fund can be summarised as (Olsen *et al.*, 2010):

- a notable reduction of serious defects in Danish social housing projects;
- the establishment of a very cost-effective insurance scheme with an insurance premium of 1 per cent compared to 2–4 per cent as the typical premium;
- the establishment of a knowledge feed-back loop mechanism between industry practitioners to improve quality of construction;
- openness on claims statistics through an open database with key figures on defects;
- the stimulation of new less-risky construction practices; and
- repair costs are estimated to have reduced by at least EUR 15 million (DKK 100 million) per year due to the dissemination of experiences.

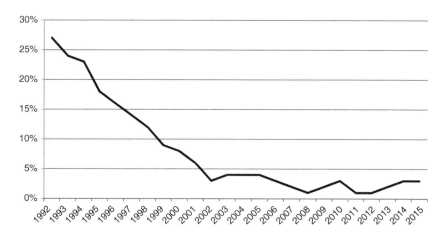

Figure 6.3 Development of serious defects in Danish social housing

Source: Adapted after Byggeskadefonden, 2015a

The fund has also published warnings about specific methods, components or materials on several occasions. Some examples include problems with structural stability of buildings, use of specific types of yellow bricks, roof tiles of cement or yellow clay and light roofing underlays (Olsen *et al.*, 2010). Lately the Danish Building Defect Fund has been instrumental in identifying wind barrier boards of MgO in wall constructions as highly unfit for use under Danish weather conditions, hence uncovering the largest building problem in decades, affecting more than 275,000 m² of walls and expected to cost around DKK1 billion, EUR 130 million, within social housing alone (Byggeskadefonden, 2015a, 2015b).

Discussion and implications for clients

In the previous sections, the characteristics of construction insurance in general were analysed followed by an analysis of the Danish Building Defects Fund as an example of an alternative way of organising construction insurance, and in particular inherent defects insurance (IDI). The analysis identifies and describes two different technological frames on construction insurance labelled 'the protection frame' and 'the driver of change frame' that governs construction practices. The comparison is summarised in Table 6.3.

Table 6.3 Two technological frames on construction insurance

	Protection frame	Driver of change frame
Goals	Transfer of risk	Improve quality of construction
Key problems	Profitable insurance schemes Pooling risks	Reduce defects Improve maintenance Financial model for repairs
Problem-solving strategies	Various market-based types of insurance dependent on local legal context: TPL/PI, CAR, IDI	Quality assurance and liability reform of legal system
Requirements to be met by problem solution	Calculating premium based on type of risks, construction lifecycle, and type of constructed assets	Self-financing scheme for social housing only
Current theories	Market economy	Negotiated economy
Design methods and criteria	Underwriting process to set premium, terms, etc.	Fixed premium of 1 per cent
Testing procedures	Risk assessment procedures Technical inspection services	Third-party independent audits/compliance checks after 1 and 5 years
Clients' and users' practices	Heterogeneous practices: all types of clients/owners	Professional and repetitive semi-public clients
Perceived substitution function	Compensation	Abolish bad construction practices
Exemplary artefacts	Policy terms and conditions Proprietary data	Open access database Guide to better quality BYG-ERFA letters

Although this chapter has focused on characterising the two different technological frames of protection and innovation rather than on the process of change as such, the analysis also points towards some important insights with regard to the process. Many different configurations of insurance schemes are available not only across the European member states, but also within the individual member states. If clients want to start pushing for new liability and insurance schemes, they would need to address at least three challenges.

The first challenge is whether a new insurance scheme should be about improving insurance as a protective mechanism or function as a driver of change in the construction industry. Hence, this question is essentially about what kind of role clients and society in general want the insurance industry to play with regard to protection as well as prevention. Consequently, the answer to this question may not only lead to a new or reformed insurance scheme in terms of premiums, covers, etc., but may also fundamentally reshape how insurance governs construction, which in turn may lead to new and more profitable construction practices.

The second challenge is about building up the technological frame to ensure that all of the elements of the technological frame are established and aligned with each other. A multitude of activities may be required in order to ensure that each of the elements is in place, for example by analysing the current situation, showcasing best practice from other fields or countries, developing various methods and guidelines, and recruiting relevant and critical actors. This process may be lengthy, cumbersome and costly, but aligning these elements is critical to achieving success on making more fundamental changes of the governing mechanisms of construction.

The third and most important challenge is to develop and strengthen a new technological frame and forge it into becoming a dominant frame. Hence, clients may want to start pushing for new liability and insurance schemes that cater for both protection of economic interests and for driving innovation in construction. At first sight this may appear fairly easy with regard to improving insurance as a protective mechanism and harder with regard to insurance as a driver of change. However, the momentum and dominance of the existing protection frame may prove very difficult indeed to change in both cases due to its deeply rooted tenets of, for example, market-driven business models, confidentiality regarding claims statistics, etc. Hence, transforming or reforming the existing liability and insurance regime may be a most daunting effort for clients to venture into, but so may the benefits.

Conclusion

This chapter has identified and described in detail two different technological frames on construction insurance, labelled 'the protection frame' and 'the innovation frame', which govern construction practices. These two technological frames have separate sets of characteristics that make them distinctively different and point towards two quite different approaches to deal with inherent defects insurance in construction. They lead to the question of what are the responsibilities and the implications for clients. It is easy for clients and client associations to be satisfied with the present state of affairs with regard to risk transfer, covers and premiums in the short term. But if clients and client associations should in some respect be dissatisfied with the present situation and the long term, this chapter has demonstrated that governing mechanisms related to construction insurance are a good place to intervene as they are malleable entities that may be changed to serve clients in other ways. Although transforming or reforming the existing liability and insurance regimes may be a most daunting effort for clients to venture into, there may be substantial benefits to reap.

References

Bijker, W. E. (1995). *Of Bicycles, Bakelites, and Bulbs: Toward a Theory of Sociotechnical Change*. Cambridge, MA: MIT Press.

Bijker, W. E. and Law, J. (eds) (1992). *Shaping Technology/Building Society: Studies in Sociotechnical Change*. Cambridge, MA: MIT Press.

Bijker, W. E., Hughes, T. and Pinch, T. (eds) (1987). *The Social Construction of Technological Systems*. Cambridge, MA: MIT Press.

Bonke, S. and Levring, P. (1996). *Building in a Market Economy: Reviewing the Danish Model*. Copenhagen: The Danish Building Development Council.

Bunni, N. G. (2011). *Risk and Insurance in Construction* (2nd edn). London: Spon Press.

BYG-ERFA (2016). *Byg på erfaringer. (in Danish: Build on experience)*. Copenhagen: BYG-ERFA. Available at: https://byg-erfa.dk/ (Accessed 9 September 2016).

Byggeskadefonden (2005). *Klassifikation af mangler ved afleveringsforretning: Foreløbig vejledning for ejere og administratorer af støttet boligbyggeri. (in Danish: Classification of Deficiencies: Preliminary Guideline for Owners and Administrators of Social Housing)*. Copenhagen: Byggeskadefonden. Available at: www.bsf.dk/media/1354/klassifikation_trykt_310805.pdf (Accessed 9 September 2016).

Byggeskadefonden (2014a). *1-års eftersyn: Forberedelse, udførelse, opfølgning. (in Danish: 1-year inspection: Preparation, Execution, Follow-Up)*. Copenhagen: Byggeskadefonden. Available at: www.bsf.dk/media/1480/1-aars-eftersyn-gaeldende.pdf (Accessed 9 September 2016).

Byggeskadefonden (2014b). *5-års eftersyn: Forberedelse, udførelse, opfølgning. (in Danish: 5-year Inspection: Preparation, Execution, Follow-Up)*. Copenhagen: Byggeskadefonden. Available at: www.bsf.dk/media/1481/5-aars-eftersyn-gaeldende.pdf (Accessed 9 September 2016).

Byggeskadefonden (2014c). *Anmeldelse af byggeskade. (in Danish: Reporting of a Building Defect)*. Copenhagen: Byggeskadefonden. Available at: www.bsf.dk/media/1482/bsf_anmeldelse_210814_low.pdf (Accessed 9 September 2016).

Byggeskadefonden (2014d). *Byggeskadefondens guide til kvalitet i boligbyggeriet* (3rd edn). *(in Danish: Danish Building Defects Fund's Guide to Better Quality in Housing Construction)*. Copenhagen: Byggeskadefonden. Available at: www.bsf.dk/media/1344/guide-3-udgave.pdf (Accessed 9 September 2016).

Byggeskadefonden (2015a). *Årsberetning 2015 (in Danish: Annual Report 2015)*. Copenhagen: Byggeskadefonden. Available at: www.bsf.dk/media/1618/bsf_beretning_2015.pdf (Accessed 9 September 2016).

Byggeskadefonden (2015b). *Byggeteknisk erfaringsformidling 2015 (in Danish: Technical Report 2015)*. Copenhagen: Byggeskadefonden. Available at: www.bsf.dk/media/1619/ bsf_erfa_2015.pdf (Accessed 9 September 2016).

Byggeskadefonden (2016a). *Byggeskadefonden (in Danish: About the Danish Building Defects Fund)*. Copenhagen: Byggeskadefonden. Available at: www.bsf.dk/ (Accessed 9 September 2016).

Byggeskadefonden (2016b). *Data byggerier. (in Danish: Data on projects)*. Copenhagen: Byggeskadefonden. Available at: www.bsf.dk/dokumentation/data-byggerier/# byggerisearch (Accessed 9 September 2016).

CEA and CSTB (2010a). *Final Report. Liability and Insurance Regimes in the Construction Sector: National Schemes and Guidelines to Stimulate Innovation and Sustainability.* Centre d'Etudes d'Assurances and Centre Scientifique et Technique du Bâtiment. Available at: www.elios-ec.eu/sites/default/files/pdf/Eliosfinalreportfullversion.pdf (Accessed 28 August 2016).

CEA and CSTB (2010b). *Special Report on Liability and Insurance Regimes in 27 EU Member States: Liability and Insurance Regimes in the Construction Sector: National Schemes and Guidelines to Stimulate Innovation and Sustainability.* Centre d'Etudes d'Assurances and Centre Scientifique et Technique du Bâtiment. Available at: www.elios-ec.eu/sites/ default/files/pdf/Eliosfinalreportfullversion.pdf (Accessed 28 August 2016).

Haugbølle, K. and Boyd, D. (2013). *Clients and Users in Construction: Research Roadmap Report*. CIB Publication 371. Rotterdam: CIB General Secretariat. Available at: http:// site.cibworld.nl/dl/publications/pub_371.pdf (Accessed 28 August 2016).

Haugbølle, K. and Boyd, D. (2016). *Clients and Users in Construction: Research Roadmap Summary*. CIB Publication 408. Rotterdam: CIB General Secretariat. Available at: http://site.cibworld.nl/dl/publications/pub_408.pdf (Accessed 28 August 2016).

Haugbølle, K. and Forman, M. (2009). Shaping Concepts, Practices and Strategies: Arbitration and Expert Appraisals on Defects. *Organization, Technology & Management in Construction: An International Journal*, Vol. 1 (1), 22–9.

Hopkin, T., Lu, S. L., Rogers, P. and Sexton, M. (2016). Detecting Defects in the UK New-Build Housing Sector: A Learning Perspective. *Construction Management and Economics*, Vol. 34 (1), 35–45.

Hwang, B. G., Thomas, S. R., Haas, C. T. and Caldas, C. H. (2009). Measuring the Impact of Rework on Construction Cost Performance. *Journal of Construction Engineering and Management*, Vol. 135 (3), 187–98.

Insurance Europe (2012). *How Insurance Works*. Brussels: Insurance Europe aisbl. Available at: www.insuranceeurope.eu/sites/default/files/attachments/How%20 insurance%20works.pdf (Accessed 28 August 2016).

Insurance Europe (2016a). *European Insurance: Key Facts. August 2016.* Brussels: Insurance Europe aisbl. Available at: www.insuranceeurope.eu/sites/default/files/attachments/ European%20Insurance%20-%20Key%20Facts%20-%20August%202016.pdf (Accessed 8 September 2016).

Insurance Europe (2016b). *Structural Data*. Brussels: Insurance Europe aisbl. Available at: www.insuranceeurope.eu/insurancedata (Accessed 8 September 2016).

Insurance Europe (2016c). *Non-life Insurance*. Brussels: Insurance Europe aisbl. Available at: www.insuranceeurope.eu/insurancedata (Accessed 8 September 2016).

Jingmond, M. and Ågren, R. (2015). Unravelling Causes of Defects in Construction. *Construction Innovation*, Vol. 15 (2), 198–218.

Josephson, P.E. and Hammarlund, Y. (1999). The Causes and Costs of Defects in Construction: A Study of Seven Building Projects. *Automation in Construction*, Vol. 8, 681–87.

Love, P. E. D. and Li, H. (2000). Quantifying the Causes and Costs of Rework in Construction. *Construction Management and Economics*, Vol. 18 (4), 479–90.

Love, P., Edwards, D. and Smith, J. (2016). Rework Causation: Emergent Theoretical Insights and Implications for Research. *Journal of Construction Engineering and Management*, Vol. 142 (6), 10.1061/(ASCE)CO.1943–7862.0001114, 04016010

Ndekugri, I., Daeche, H. and Zhou, D. (2013). The Project Insurance Option in Infrastructure Procurement. *Engineering, Construction and Architectural Management*, Vol. 20 (3), 267–89.

Nielsen, G. (2012). *Byggeskadefonden 1986–2011: De første 25 år. (In Danish: Danish Building Defects Fund 1986–2011: The first 25 years.* Copenhagen: Byggeskadefonden. Available at: http://byggeskadefonden.dk/media/1427/bsf_25_aar.pdf (Accessed 28 August 2016).

Nielsen, J., Pedersen, C. and Hansen, M. H. (2004). *Svigt i byggeriet: Økonomiske konsekvenser og muligheder for en reduktion (in Danish: Defects in construction: Costs and possible reductions)*. Copenhagen: Erhvervs-og Byggestyrelsen.

Olsen, I. S., Bertelsen, N. H., Frandsen, A. K. and Haugbølle, K. (2010). *Defects in Housing, Musikbyen. Danish Building Defects Fund (BSF). CREDIT case DK08. SBi 2010:27.* Hørsholm: Danish Building Research Institute, Aalborg University.

Oudshoorn, N. and Pinch, T. (eds.) (2003). *How Users Matter: The Co-Construction of Users and Technologies.* Cambridge, MA: The MIT Press.

Pedersen, O. K., Andersen, N. Å., Kjær, P. and Elberg, J. (1992). *Privat politik: Projekt Forhandlingsøkonomi. (in Danish: Private Politics: Project Negotiated Economy)*. Copenhagen: Samfundslitteratur.

Pedro, J. B., Meijer, F. and Visscher, H. (2010). Building Control Systems of European Union Countries. *International Journal of Law in the Built Environment*, Vol. 2 (1), 45–59.

Ranasinghe, M. (1998). Risk Management in the Insurance Industry: Insights for the Engineering Construction Industry. *Construction Management & Economics*, Vol. 16 (1), 31–9.

Roussel, J., Droogenbroek, M., Van, Salagnac, J.-L., Vermande, H., Dunand, T. and Haugbølle, K. (2015). *ELIOS Final Report.* Brussels: CEA Belgium. Available at: www.elios-ec.eu/sites/default/files/elios2-final-report.pdf (Accessed 28 August 2016).

Scanlon, K. and Vestergaard, H. (2007). The Solution, or Part of the Problem? Social Housing in Transition: The Danish Case. In: *European Network of Housing Research, ENHR Proceedings, Sustainable Urban Areas, International Conference, 25–28 June, Rotterdam, 2007.* Available at: http://eprints.lse.ac.uk/29945/1/The_solution_or_part_of_the_problem_(author).pdf (Accessed 9 September 2016).

Sun, M. and Meng, X. (2009). Taxonomy for Change Causes and Effects in Construction Projects. *International Journal of Project Management*, Vol. 27 (6), 560–72.

Sudres, R. (2011). *Construction Risk Insurance.* Hanover: Hannover Re.

7 Construction management capabilities of clients

A methodology for assessment

Youngsoo Jung and Seunghee Kang

Introduction

A client decides and influences the primary direction of a construction project, and one of the client's important decisions in the early planning phase is to select appropriate project delivery methods (PDM) so as to outsource engineering and construction services in an optimised manner. Selected project delivery methods (PDM) well represent distinct managerial requirements of the client. In this sense, the CIB research roadmap of Working Commission W118 (Haugbølle and Boyd, 2013) pointed out research issues in the contents and scale of clients' and user's value chains in various national and institutional contexts. This roadmap also suggested developing a coherent model of what constitutes a client and a user under different structural conditions. In an attempt to quantitatively represent such a coherent model, this chapter introduces a methodology for assessing the construction management capability of clients. Implications from Korean clients are also presented in order to illustrate the methodology.

One of the early studies analysing project delivery methods (PDM) in a quantitative manner was by Konchar and Sanvido (1998). In their study, data collected from 351 real-world projects were thoroughly examined in order to statistically compare the benefits of different types of PDM including design–bid–build (DBB), design–build (DB) and CM at Risk (CMR). Koppinen and Lahdenperä (2007) added design–build–maintain (DBM) in addition to DBB, DB and CMR comparison. Koppinen and Lahdenperä (2007) performed an analysis with a different method by comparing the benefits of four different PDMs for single model project. Therefore, it was possible to illustrate cost savings and schedule reductions in monetary amount and working days, respectively. Interestingly, a recent study by El Asmar *et al.* (2013) added integrated project delivery (IPD) into this type of analysis. Although the methodologies and scopes were somewhat different, the three studies well developed metrics for performance assessment (mainly including cost, time and quality) in order to compare advantages and drawbacks of different PDMs from the client's perspective. These studies imply that alternative PDM including DB, CMR and IPD significantly outperform the traditional DBB, and this result supports the reason why alternative PDM is becoming prevalent in the global construction industry. Another important notion is that clients' involvement and capability is a crucial factor under the alternative PDM contracts.

Despite active efforts being made to explore alternative PDM over the past two decades, there have not been enough studies conducted for the client organisation itself, which plays a significant role in successful construction projects. Even existing studies have rarely discussed in-house construction management (CM) capability based on clients' construction business functions. Different types of the client organisation may require different levels of client involvement in the construction process. This fact may provide insights to explore the clients and users value chain defined by Haugbølle and Boyd (2013). Therefore, understanding a client's CM capability facilitates efficient planning of new construction projects for clients and for architectural, engineering and construction (AEC) service providers. In this context, the purpose of this chapter is to explore the variations of the clients' CM capability based on comprehensive construction business functions.

The CM function of a client could be only a small part of the entire client functions as the client's major business would not be in the construction industry. As depicted in Figure 7.1, different characteristics of clients (box 1 – 'Client Type') determine the form of clients' project management organisation (PMO) in terms of weighting, depth and capability of construction business functions (box 2) based on the client's business area. In turn, the client PMO directly influences AEC companies' roles and responsibilities (box 3). Though this paper focuses on the client type and its CM capability (boxes 1 and 2), relevant issues such as PDM, market statistics and communication requirements are also briefly discussed by introducing several studies conducted at Myongji University in South Korea.

With this model in mind, the objectives of this chapter are:

- to develop an evaluation methodology for measuring the construction management capabilities of different types of client organisation based on fourteen construction business functions; and
- to demonstrate the applicability of the methodology in an exploratory study to assess the gap between current and required construction management capabilities of Korean public clients.

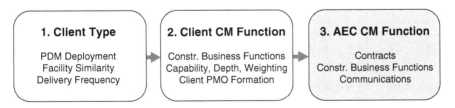

Figure 7.1 Construction management (CM) functions of clients

The structure of this chapter is as follows:

- First, this chapter will introduce a theoretical framework to provide a typology of clients and identify fourteen business functions derived from the PMBOK by the Project Management Institute.
- Second, this chapter will develop the evaluation methodology to analyse the capability gap between required and current capabilities along with the preferences for conducting the business function in-house or out-sourced.
- Third, this chapter will apply the evaluation methodology in an exploratory study of Korean public construction clients.
- Finally, this chapter will emphasise the value for construction clients of applying a more systematic approach to the assessment of their own construction management capabilities and point at the importance for setting up appropriate collaboration with the construction service providers in the AEC industry.

Theoretical framework: client types, business functions and capabilities

In order to characterise client organisations from a construction perspective, four different types of client are first defined based on two variables of clients' construction projects: *delivery frequency*, and *facility similarity*, as shown in Figure 7.2. For example, a knowledgeable client may execute repeating projects of a similar type of facilities every year. Secondly, an assessment methodology for clients' CM capability is also developed to understand their in-house CM capability based on fourteen different construction business functions defined by Jung and Gibson (1999). Based on these definitions and methodologies, surveys and case-studies were performed in order to analyse industry statistics (Jung *et al.*, 2014; Kang *et al.*, 2014), to examine client characteristics (Jung *et al.*, 2004) and to explore information exchange between project participants (Jung *et al.*, 2011) for the sake of better collaboration between clients and AEC service providers.

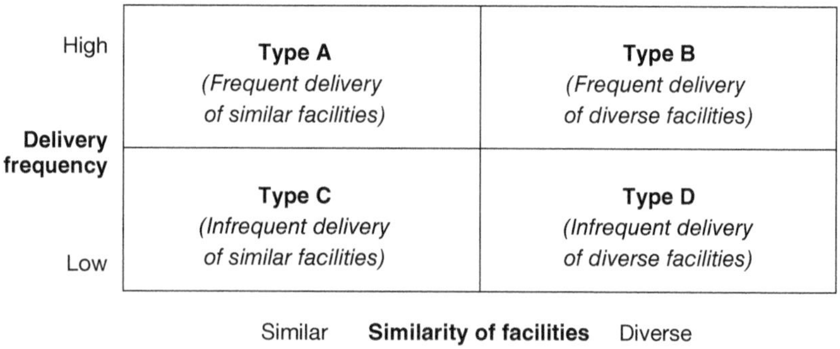

	Similar	Diverse
High ↑ Delivery frequency ↓ **Low**	**Type A** *(Frequent delivery of similar facilities)*	**Type B** *(Frequent delivery of diverse facilities)*
	Type C *(Infrequent delivery of similar facilities)*	**Type D** *(Infrequent delivery of diverse facilities)*

Similar   **Similarity of facilities**   Diverse

Figure 7.2 Four types of client organisation

Source: Jung *et al.*, 2004: 129

Four different types of client

Clients for construction projects vary widely according to their organisational nature, characteristics and purposes. For instance, there are organisations in which their purpose of establishment is closely associated with the construction industry, whereas there are many others in which the role of being a construction project client is merely a minor duty. For the latter cases, frequency or scale of construction projects accounts for a very small part of their entire business efforts. Eventually, this diversity in client organisations demonstrates various requirements for in-house CM method and demand. Therefore, the client organisations must be fully aware of their characteristics and identify their own CM capabilities. Based on the identified requirements, a client can determine managerial policy for its construction projects including strengthening of their own capabilities or outsourcing to external organisations.

From this perspective, Jung *et al.* (2004) introduce 'delivery frequency' and 'facility similarity' as being two variables that categorise different types of client organisations into four different quadrants as illustrated in Figure 7.2. This classification has been employed as a framework for performing case studies and also for deriving implications. The four types of clients are:

- Type A: The upper-left quadrant identifies clients who frequently build similar facilities, where specialised in-house CM organisation and manpower is highly required. For 'Type A' clients, CM capabilities can be strengthened in-house to enable efficient project management throughout repetitive similar projects.
- Type B: The clients in the upper-right quadrant also frequently build facilities. However, the type of built facilities is so diverse that it is difficult for Type B clients to have in-house speciality based on repeating experiences (e.g. in-house design management).
- Type C: Type C clients in the lower-left quadrant build similar facilities infrequently, thus having less interest in technical issues of CM. Type C clients may want to have better planning capability for their own construction projects.
- Type D: In the lower-right quadrant Type D clients infrequently deliver diverse types of facilities. Type D clients may include a special-purpose company (SPC) or a temporary consortium of many organisations for a joint effort in delivering a facility.

Fourteen construction business functions

Different types of client may have different emphasis in their project management. For the purpose of specifying these managerial requirements, it is useful to define practical variables required to perform construction projects. In this sense, CM requirements can be clearly presented by using construction business functions (e.g. scheduling and cost control), even though there are many other measures to indicate construction activities, including project life cycle, building types and management cost. Again, the construction business functions can be defined in different levels or perspectives. As a good example, the project management handbook PMBOK by PMI (1996) classifies knowledge areas into nine categories which are integrated management, service management, schedule control, project cost control, personnel management, information management, risk management and contract management. However, the functions categorised by PMBOK provide a universal classification applicable to any type of industry including automobile, information and communication technology (ICT) and defence. Therefore, it might not be fully construction-specific. Another notion is that, in practice, it is widely conceived that less than five functions are not detailed enough while over twenty functions require tremendous efforts to classify and analyse in an intuitive and fairly easy manner.

After comparing existing classifications of management functions (e.g. PMI, 1996; CII, 1990) and by incorporating industry-specific characteristics of the construction industry, this study used fourteen different business functions defined for construction projects (Jung and Gibson, 1999):

- planning (B01);
- sales (B02);
- design (B03);
- estimating (B04);
- scheduling (B05);
- materials management (B06);
- contracting (B07);
- cost control (B08);
- quality management (B09);
- safety management (B10);
- human resource management (B11);
- finance/accounting (B12);
- general administration (B13); and
- research and development (B14).

These functions can be further decomposed into sub-functions in order to better define each organisation's managerial requirements as exemplified in Table 7.1.

Formation of construction management capabilities

Types of organisation and business function were previously discussed to characterise the clients. The next issue for assessment variables is the formation of CM capability, which concerns how a client physically focuses on its CM capability inside its organisation. Namely, a client may want to have:

- a specialised in-house department for construction management (labelled 'organisational' under the 'Formation' variable in Table 7.2);

Table 7.1 Fourteen construction business functions

Business functions	Examples of sub-functions for the business functions
(B01) Planning	Feasibility study, Project execution plan, Project charter
(B02) Sales	Project development, Bid planning, Project contracting
(B03) Design management	Preliminary design, Detailed design, Shop drawing, Specifications
(B04) Estimating	Design estimate, Bid estimate, Control estimate
(B05) Scheduling	Planning and scheduling, Construction planning
(B06) Materials management	Materials management, Equipment management
(B07) Contracting	Subcontracting, Subcontractor training and appraising, Dispute resolution
(B08) Cost control	Budgeting, Cost monitoring, Forecasting
(B09) Quality management	Quality control, Quality assurance, Aftersales services
(B10) Safety management (HSE)	Pre-appraisal, Prevention, Education, Settlement
(B11) Human resource management	Employment, Posting, Wages, Fringe benefits, Education, Corporate culture
(B12) Finance/accounting	Financing, Managerial accounting, Credit management, Securities
(B13) General administration	General affairs, Auditing, Legal affairs, Facility and document management
(B14) Research and development	Research and development, Technology management, Information systems

Source: Adapted after Jung and Gibson, 1999: 218

- to hire more in-house engineers (labelled 'manpower' under the 'Formation' variable in Table 7.2); or
- to attain high level CM knowledge and skills (labelled 'technical' under the 'Formation' variable in Table 7.2) in order to oversee service deliveries from AEC companies.

These three types of formation – organisational, manpower and technical methods – can be applied together. *Organisational formation* represents a systematic approach in clients' CM functions, especially in a large organisation. *Manpower formation* focuses on the weightings of required man-hours, where a client wants to handle CM functions by himself. Finally, *Technical formation* indicates the superior level of CM knowledge.

Table 7.2 Assessment variables for client CM capability

Subject	Variable	Value/Description
Client type	D: Delivery frequency	High Low
	S: Similarity of facilities	Similar Diverse
Client CM capability	B: Construction business functions	Planning (B01) Sales (B02) Design (B03) Estimating (B04) Scheduling (B05) Materials management (B06) Contracting (B07) Cost control (B08) Quality management (B09) Safety management (B10) Human resource management (B11) Finance/accounting (B12) General administration (B13) Research and development (B14)
	F: Formation	Organisational Manpower Technical
	G: Demand	To-Be As-Is Gap
	M: Enhancement method	In-house Outsourcing

Source: Adapted after Jung *et al.*, 2004a: 130; and Jung and Gibson, 1999: 218

In addition to construction business functions and formation, a third perspective on demand was also used to assess client CM capabilities. This variable of 'demand' efficiently locates the areas for improvement and future plans. The 'demand' variable includes three statuses:

- To-Be: analysis of required capabilities for implementation in near future;
- As-Is: evaluation of present capabilities; and
- Gap: understanding the lack of capabilities.

Finally, managing and acquiring the required capabilities can be further categorised into in-house and/or outsourcing as the *Enhancement method*. In other words, if it appears necessary that an organisation possesses specialised CM organisations and capabilities, it can do so by employing new manpower or training internal personnel. Or, where outsourcing a CM organisation becomes requisite due to certain management conditions, the scope and the details of outsourcing can be comprehended.

In summary, the in-house construction management (CM) capability varies by the type of client organisation that can be characterised based on 'delivery frequency' and 'similarity of facilities'. The CM requirements for different client types can be further specified by using 'construction business functions', organisational 'formation', practical 'demand' and 'enhancement method' to acquire the required capabilities.

Evaluation methodology for client CM capabilities

For the purpose of developing an evaluation methodology for the client's CM capability, the assessment variables are defined first. Two subjects (i.e. client type and client CM capability) with the six variables summarised in Table 7.2:

- D: delivery frequency;
- S: project similarity;
- B: construction business functions;
- F: formation;
- G: demand; and
- M: enhancement method.

By using these variables, Jung *et al.* (2004) developed an assessment methodology for clients' CM capability as depicted in Table 7.3, which focuses on quantifying the relative importance of each construction business function in order to indicate overall insights of CM capability for each type of client. Therefore, fourteen construction business functions by Jung and Gibson (1999) were used as the main measure (dotted box B in Table 7.3). Detailed sub-functions of each business function can be further defined to incorporate its own organisational specifics. For example, the case-study of this chapter used the official 'Construction Management Implementation Guidelines by Korean Ministry of Construction and Transportation' (MOCT, 2001) to describe the client's functions in a practitioner-friendly manner.

As for the formation (F) variable in Table 7.2, three different formations of CM capability (organisational, manpower, technical) are individually evaluated to differentiate the forms of CM requirements. Dotted box 'F' in Table 7.3 depicts how the formation is used in the assessment. In order to calculate the capability gap (future required vs. current capabilities), the variable of 'demand' was used (dotted box 'G' in Table 7.3). Finally, the preference to in-house or outsourcing for attaining capability gaps is quantified in the last two columns of Table 7.3 (dotted box 'M'). The value for 'preference in-house' indicates the respondents' preference to have enhanced CM capability in-house or outsourced. The lower value of 'preference for in-house' indicates the client's desire to outsource that specific business function. This measure is used in order to quantitatively represent the differences of in-house preferences among the 14 different business functions.

Next part of the proposed assessment methodology is to set up quantifying mechanisms. Quantifying each variable in the study utilises normalised scores for

Table 7.3 Assessment methodology for client CM capabilities

CONSTRUCTION BUSINESS FUNCTION	REQUIRED CAPABILITY (TO-BE)			CURRENT CAPABILITY (AS-IS)			CAPABILITY GAP (TO-BE VS. AS-IS) G			PREFERENCE FOR IN-HOUSE*	
	Organisation	Manpower	Technology	Organisation	Manpower	Technology	Organisation	Manpower	Technology	Manpower	Technology M
Planning		5									
Sales		4									
Design management		2									
Estimating		2									
Scheduling		3									
Materials management		2									
Contracting		1									
Cost control		2									
Quality management		2									
Safety management		2									
Human resource mgmt.		2									
Finance/accounting		2									
General administration		2									
Research and development		4									2
Total		35									1400

Source: Adapted after Jung *et al.*, 2004a: 131

Note: * 'The preference value for in-house' indicates the respondents' preference to have enhanced CM capability in-house or outsourced.

each construction business functions (B) as listed in the column box '1' in Table 7.3. For example, let's assume that an assessment used a scale of 1 to 5 to evaluate each business function. We can also assume that the score for 'required (To-Be)' 'man-power' for 'sales' function was 4 out of 5, where column total of 'required manpower' was 35 out of 70 (i.e. maximum 5 points for each of the 14 functions). Note that the scores might be an average from many respondents. The score (e.g. 4 for sales function) for each business function is then normalised by comparing it with total scores of all of fourteen business functions (e.g. 35). Since the normalised score is a ratio of one business function to all others, it can represent the relative degree of importance where the score of 100 means exact average and median. Therefore, the normalised score for 'required (To-Be)' 'man-power' for the 'sales' function is 160 (4/35 x 14 x 100). Box 1 shows original scores for this example. Note that the sum of normalised scores for each column should be 1,400.

By repeating this process, each business function can be evaluated in terms of organisation, manpower and technology (i.e. 'formation', as shown in the box 'F'). Also, the same process needs to be performed again in order to evaluate current capability as well as required capability. The capability gap can then be automatically calculated (in box 'G'). The final step is to evaluate the preference in capability enhancement method (box 'M') by using the same normalising technique.

Lessons from an exploratory study of Korean public clients

For a better understanding of the application of the survey developed in this chapter, we will introduce a brief overview of Korean construction market. At the end of 2012, total investment for domestic construction was about USD 150 billion in South Korea, while revenue of Korean construction organisations from the international construction market in the same year was about USD 65 billion, making the total size of the Korean construction industry USD 215 billion. Half of the domestic construction investment is for public projects; the other half is in the private sector.

Though the basic industry characteristics for building, civil infrastructure and industrial plant construction are similar to those in other countries, the housing market in Korea has distinct characteristics. The major form of housing in Korea is the high-rise apartment complex, forming a fairly big community, so, due to their huge project size, many housing projects are attractive to large AEC firms. Most homes are individually owned, and a limited number of housing is available for lease by public or private facility developers, which makes for a unique situation where houses are often leased by individual home-owners who have less than three houses. Many people prefer to invest in housing as Korea has experienced continuously rising housing prices over several decades, even though it is not that much prevalent any more since the global financial crisis. In order to prohibit excessive booms in the housing market, ownership of more than three homes by one person is subject to higher property tax unless owned and operated by a corporate body.

In the following section the application of the developed methodology for assessing clients' CM capabilities will be demonstrated for two separate yet interlinked purposes:

- First, the methodology will be used to analyse the CM capabilities of one client or a group of clients. This may be relevant for clients wishing to analyse his/her own CM capabilities in order to identify areas for improvement.
- Second, the methodology will be used to analyse the CM capabilities of clients being grouped into different types. This may be relevant for conducting an industry analysis to formulate public policies or for a client association that wants to develop new educational activities for various target groups.

Capability analysis of clients' CM functions

A survey questionnaire was developed and sent out to Korean public clients in 2003 (Jung et al., 2004) by using the assessment methodology for clients' CM capabilities. Organisations surveyed include publicly owned companies, local governments and public companies, of which 43 organisations were selected randomly, with 14 of them providing responses indicating a response rate of 33 per cent. With the exception of 3 of the above responses, which included partially inadequate contents, 11 of them were applied to evaluate client organisations (five respondents of client Type A, two of Type B, four of Type C and none of Type D). As the number of respondents was limited, it was difficult to grant statistical significance to the survey results, but the study can be interpreted as a test of the methodology in practice.

In the questionnaire to the respondents, a Likert scale of 1 to 5 was used to evaluate required capability and current capability in terms of organisational, manpower and technical formation in each organisation's area of CM. For example, for required capability a score of 1 signifies 'not required' and a score of 5 signifies 'highly required'. Therefore, by subtracting the current capability value from the required capability value, the lack of capability was computed. The average score of one function was then normalised by comparing it with total scores of all 14 business functions in order to show relative percentile importance of one business function to all others. Therefore, the score of 100 indicates exact median and average. After normalisation, if the required capability (or present capability) value goes higher than a score of 100, it means that the capability of that specific function is relatively more required than other functions.

Table 7.4 presents mean values for the 11 public Korean client organisations. Construction business functions with highly required capability (To-Be) in manpower formation were found to be in the order of planning (125), estimation (115), R&D (113, mainly information systems and manuals for construction) and design management (110). Business functions with a significant capability gap, which is computed by subtracting present manpower capability from required manpower capability, appeared in the order of planning (28), estimation (11) and design management (11). Planning, estimating and design management were found to be the most important CM functions for clients. Based on the survey, it was also found that Korean clients have a preference for outsourcing for design management (83), safety management (87), scheduling (87) and quality management (89) in terms of CM manpower. The values for manpower in Table 7.4 are highlighted for easier locating. As for the preference for in-house measure in Table 7.4, a score below 100 indicates that the client prefers to outsource, while a score above 100 means there is preference for in-house. This normalised score is relative for 1 among 14 different business functions.

Though the survey was evaluated in terms of three measures – organisation, manpower and technology – for the formation variable as defined in Table 7.2, the measure of manpower was discussed above as an illustration. Other measures can be used together for the purpose of self-evaluation of a client CM capability to provide meaningful insight. For example, in Table 7.4, the score of required capability in terms of organisation for the finance/accounting function is 111, which is higher than the score of required capability in terms of manpower for the same function (99). This implies, again relatively among 14 functions, that they do not want to have more manpower for the finance/accounting business function.

Table 7.4 Capability analysis of clients' CM functions

Construction Business Function	Required Capability (To-Be) Current			Capability (As-Is)			Capability Gap (To-Be vs. As-Is)			Preference for In-house	
	Organisation	Manpower	Technology	Organisation	Manpower	Technology	Organisation	Manpower	Technology	Manpower	Technology
Planning	122	125	120	99	97	104	23	28	16	98	99
Sales	95	93	94	92	86	85	3	7	9	95	85
Design management	115	110	112	99	99	106	16	11	6	83	79
Estimating	116	115	108	104	104	111	12	11	-3	104	105
Scheduling	102	104	104	108	108	112	-6	-4	-8	87	88
Materials management	75	84	89	98	105	98	-23	-21	-9	101	101
Contracting	101	106	109	109	115	113	-8	-9	-4	114	105
Cost control	89	99	94	92	98	96	-3	1	-2	94	106
Quality management	94	97	103	99	99	98	-5	-2	5	89	86
Safety management	87	88	83	97	97	92	-10	-9	-9	87	92
Human resource mgmt.	82	84	87	95	88	93	-13	-4	-6	109	118
Finance/accounting	111	99	101	104	102	97	7	-3	4	118	118
General administration	94	83	88	95	92	85	-1	-9	3	113	114
Research and development	117	113	108	109	110	110	8	3	-2	108	104
Total	1,400	1,400	1,400	1,400	1,400	1,400	–	–	–	1,400	1,400

Source: Jung et al., 2004a: 131

However, they would like to set up a more consolidated organisational structure for finance/accounting function.

Another example for elaborating the technology formation could be the function of quality management. The scores of required capability for organisation, manpower and technology are 94, 97 and 103, respectively. It reveals that the Korean clients want to have advanced skill and technical capability in the near future, rather than having more people for quality management function. By assessing these three formations, of organisation, manpower and technology, in this way, a client organisation can systematically evaluate its future direction of CM capability.

Capability analysis across different types of clients

In order to graphically see the different CM requirements among different 'Client Types', Table 7.5 summarises the survey result (Jung *et al.*, 2004) by groups. As there were no responses from Type D, this category has been left out for simplicity. In Table 7.5, required capability and capability gap are symbolised by using circles and triangles based on normalised scores, where:

- the symbol ● means relatively high and ○ is relatively low in 'required capability';
- the symbol ▲ is relatively high (insufficient) where △ is relatively low (sufficient) for capability gap; and
- the solid ones (● and ▲) are used to emphasise those functions that are highly demanding functions (relatively higher capability required) and those that have bigger gaps (currently insufficient ones).

Type A includes clients which deliver similar projects frequently (e.g. public housing agency), where necessity for their own specialised CM organisation and manpower is relatively high. As part of the survey results, from the manpower perspective for Type A, CM functions with highly required capability are planning (127), estimation (124), contracting (124) and design management (114). For the current capability, it appears in the order of contract management (129), R&D (120) and scheduling (113). In particular, planning, design management and estimating were chosen as being the most significant To-Be CM functions (● in Table 7.5 under column title of R1) as well having the most insufficient manpower (▲ in Table 7.5 under column title of G1).

Type B clients deliver diverse projects frequently, where their own specialised CM organisation and manpower are required. At the same time demand for outsourced CM experts also exists to a certain degree. As for organisational formation, Type B requires systematic in-house organisations for design management (127), estimating (127) and planning (108) functions as current capabilities do not meet the required capabilities for those three functions. As for the quality management function, responses were made as having possessed more than required. It is interesting that clients tend to focus less on job site supervision by their own employees. This fact has also been supported by a recent survey by

Table 7.5 Areas for improvement of CM capability for different types of clients

Construction Business Function	Required Capability — Client Type A — R1 Manpower	Required — Client Type A — R2 Technology	Required — Client Type B — R3 Manpower	Required — Client Type B — R4 Technology	Required — Client Type C — R5 Manpower	Required — Client Type C — R6 Technology	Capability Gap — Client Type A — G1 Manpower	Gap — Client Type A — G2 Technology	Gap — Client Type B — G3 Manpower	Gap — Client Type B — G4 Technology	Gap — Client Type C — G5 Manpower	Gap — Client Type C — G6 Technology
Planning	●	●	O		●	●	▲			▲	▲	▲
Sales	O	O	O			●			△		▲	▲
Design management	●		●	●	O		▲	△	▲	▲	▲	▲
Estimating	●		●	●			▲		△	△	▲	△
Scheduling							△	△				△
Materials management	O				O	O	△				△	△
Contracting	●	●	O		O	●	△				△	△
Cost control	O	O	O			O		▲				▲
Quality management	O	O	O	O		O	△				△	△
Safety management	O	O	O	O		O	▲				△	△
Human resource management			O	O		O	▲	▲			△	△
Finance/accounting		O	O	O	O		▲	▲			△	△
General administrations	O	O	O	O			△	△		△	△	△
Research and development		●	●		●						▲	▲

Source: Adapted after Jung et al., 2004a: 134

Legend:
Required Capability: ● Relatively High O Relatively Low
Capability Gap: ▲ Relatively High (Insufficient) △ Relatively Low (Sufficient)
Blanks have relatively median value for required capability or lacking capability (i.e. approximate value of 100 for required capability, approximate value of 0 for lacking capability).

Jung *et al.* (2014). Type B clients are interested in enhancing CM capabilities for design management and estimating in all formations (organisational, manpower and technical). However, they prefer to outsource design management.

Type C clients build similar types of facility infrequently. One of the major findings is that Type C clients are quite satisfied with their current CM capabilities. They may hire outside CM consultants whenever needed for effectiveness. Nevertheless, they still desire to significantly enhance planning capabilities in terms of manpower and technical formation (under the column title of R5, R6, G5, G6 in Table 7.5). They are also interested in improving R&D (information systems and manuals), sales, design management and cost control.

Type D clients provide infrequent delivery of diverse facilities. As this study of Korean public clients did not achieve responses for this particular type of client, no further discussion will be applicable here. It is hoped that future studies will bring provide additional information on this type of clients.

In summary, the CM functions of planning, design management, estimating and R&D (systemisation and manuals) are of significance to all types of client as shown in Tables 7.4 and 7.5. Among those four functions, planning has the highest priority for improvement in terms of organisational, manpower and technical formations. In particular, emphasis has been put on planning in Type A and Type C, both of which repeatedly build similar types of facility. As for design management, lack of manpower in particular was indicated in all client types. Capability enhancement for estimating is emphasised by frequently delivering clients of Type A and B. Figure 7.3 outlines the survey results and implications from Korean clients.

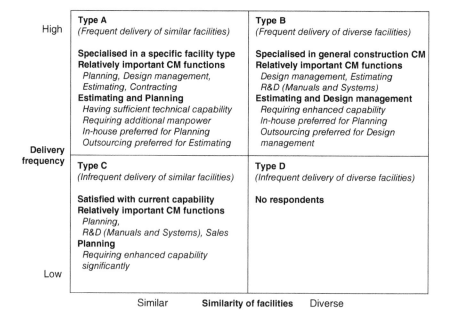

Figure 7.3 Summary of the Korean case study

Conclusion

Selecting a project delivery method is a construction client's most important decision in the early phase of a project. Assessing and understanding a client's own construction management functions, therefore, can facilitate effective selection of an appropriate project delivery method that shares and distributes roles and responsibilities with AEC companies in the most efficient manner. The need to understand the client's own construction management capabilities is becoming even more pressing as alternative project delivery methods including design–build, public–private partnerships, construction management at risk and integrated project delivery are currently becoming more prevalent.

Notwithstanding the utmost significance of the client's role in realising project objectives for construction, researchers have rarely explored the construction management functions of construction clients. In order to address this issue, this chapter has defined four types of client, with the purpose of evaluating the construction management capabilities of client organisations, and contemplated CM demands for each client type. With these challenges in mind, this chapter has developed an evaluation methodology for measuring the construction management capabilities of different types of client organisation based on 14 construction business functions, and it has demonstrated the applicability of the methodology in an exploratory study of the gap between current and required construction management capabilities among Korean public clients.

The applicability of the methodology has been demonstrated for two separate yet interlinked purposes:

• The methodology has been applied to analyse the CM capabilities of a single client or group of clients wishing to analyse their own CM capabilities in order to identify areas for improvement.
• The methodology has been applied to analyse the differences in CM capabilities between clients being grouped into different categories. This may be relevant for conducting an industry analysis to formulate public policies or for a client association that wants to develop new educational activities for various target groups.

This chapter has attempted to develop an easy-to-use methodology for comprehensively evaluating a client's in-house construction management capabilities. The development of the methodology has been based on generally acceptable variables for construction management practice as reflected in 14 business functions derived from the general project management literature. Therefore, even though this chapter introduces the nation-specific case of Korea, the methods and variables are expected to be applicable universally regardless of regions, regulations and market conditions.

References

CII (1990). *Assessment of Construction Contractor Project Management Practices and Performance: A Special Publication of Construction Industry Institute (CII)*. Austin, TX: University of Texas.

El Asmar, M., Hanna, A. and Loh, W. (2013). Quantifying Performance for the Integrated Project Delivery System as Compared to Established Delivery Systems. *Journal of Construction Engineering Management*, Vol. 139 (11), 04013012.

Haugbølle, K. and Boyd, D. (2013). *Clients and Users in Construction: Research Roadmap Report*. CIB Publication 371. Rotterdam: International Council for Research and Innovation in Building and Construction (CIB).

Jung, Y. and Gibson, G. E. (1999). Planning for Computer Integrated Construction. *Journal of Computing in Civil Engineering*, Vol. 13 (4), 217–25.

Jung, Y., Joo, M. and Kim, H. (2011). Project Management Information Systems for Construction Managers: Current Constituents and Future Extensions. In: *Proceedings of the 28th International Symposium on Automation and Robotics in Construction.*, Seoul: ISARC, pp. 597–602.

Jung, Y., Shin, D., Kang, S. and Kim, N. (2014). Growth Model for Korean CM Firms based on 2012 Statistics, *Korean Journal of Construction Engineering and Management*, Vol. 15 (6), 92–104.

Jung, Y., Woo, S., Park, J., Kang, S., Lee, Y. and Lee, B.-N. (2004). Evaluation of the Owners' CM Functions, *Korean Journal of Construction Engineering and Management*, Vol. 5 (3), 128–36.

Kang, S., Jung, Y., Kim, N. and Shin, D. (2014). Policies and Tasks for Improving Korean CM Industry, *Korean Journal of Construction Engineering and Management*, Vol. 15 (5), 71–81.

Konchar, M. and Sanvido, V. (1998). Comparison of US Project Delivery Systems. *Journal of Construction Engineering Management*, Vol 124 (6), 435–44.

Koppinen, T. and Lahdenperä, P. (2007). Realized Economic Efficiency of Road Project Delivery Systems. *Journal of Infrastructure Systems*, Vol. 13 (4), 321–9.

MOCT (2001). *A Guide to Construction Management Practices, Notification of the Korean Ministry of Construction and Transportation (MOCT)*, Seoul, Korea.

PMI (1996). *PMBOK: A Guide to Project Management Body of Knowledge*. Upper Darby, PA: Project Management Institute.

8 Client learning across major infrastructure projects

Leentje Volker and Mieke Hoezen

Introduction

As initiators of projects and as significant contracting agencies, clients are important actors in the construction industry (Atkin *et al.*, 1995). By engaging directly in the planning and construction of new projects, they not only shape the product, but also the construction process (Hartmann *et al.*, 2008). Recent research indicates that formal aspects of the legal contract, informal aspects of the relation between contracting parties and the involvement of stakeholders are of great influence to the governance of projects (Eriksson and Westerberg, 2011; Lizarralde *et al.*, 2013).

The construction industry is showing a tendency for clients to involve their contractors in projects earlier. The contracts to govern construction projects are also signed earlier in the process, when client ideas are not yet fully developed and when the risk of unforeseeable contingencies are considerable. This makes both clients and contractors feel the need to have conversations before a contract is signed (Dorée, 2001). Thus, they are able to come to a better understanding on project details, the allocation of risks and the terms for cooperation. These aspects, which are of great influence to project success (Tabish and Jha, 2011), are discussed and negotiated during the procurement stage of projects.

In previous research, we demonstrated how formal and informal commitment simultaneously develop and grow during the procurement stage of projects (Hoezen, 2012; Hoezen and Volker, 2015; Volker, 2012). In this chapter, we explore how procuring agencies can learn from their own projects by studying the decision-making process in previous projects. This contributes to the W118 research and development agenda (Haugbølle and Boyd, 2013) by deepening the understanding of mechanisms behind the regulation of supply with the construction industry. The work focuses on elements influencing project success and reasons why certain client behaviour in procurement situations may be more effective than others. It, therefore, contributes to the understanding of governance mechanisms of client organisations. We describe how the learning experience of the Dutch Highway Agency (in Dutch: Rijkswaterstaat) led to adaptation of project governance structures in three consecutive infrastructure projects.

Theoretical background

Project governance

Project governance concerns how a project is formally designed and controlled. Literature on project governance appears to focus on two different areas: governing projects in organisations (see e.g. Hobday, 2000; Ruuska *et al.*, 2011; Thiry and Deguire, 2007), and governance of inter-organisational projects (e.g. Bosch-Sijtsema and Postma, 2010; Caldwell *et al.*, 2009). In this research we focus on learning processes that relate to the governance of projects in organisations. Inspired by the structure–conduct–performance paradigm for strategic management (Mason, 1939; McWilliams and Smart, 1993), we identify structure, people and information as the three main pillars of project governance.

Commitment

Ring and van de Ven (1994) indicate that how projects are governed is established and codified in a formal legal contract, and informally understood in a psychological contract between the parties. They conclude that the commitment stage of projects – during which those contracts are formed – is of major importance to the development of projects. Hellström *et al.* (2013: 712) explain this by their finding that '*strong relationships and commitment open up opportunities for creating alternative paths during project appraisal, increase the array of available governance mechanisms, and hence lay the foundations for the final governance structure of the project execution phase*'. This is further stressed by Eriksson and Westerberg (2011), who argue that projects managed within cooperative relationships are more successful than other projects. It is therefore essential for the learning capacity of an organisation to identify the lessons learned during the development of commitment and project governance under the influence of procurement processes.

   The commitment of the client organisation and its contractor to the project forms the basis for cooperation between them (Ring and van de Ven, 1994). Legally, commitment often refers to the state of being bound to a course of action or to another party. It stems from both natural motivators as feelings of empathy or shared values, and from artificial motivators such as contract clauses and reward mechanisms. Commitment to the project by both the public client and the contractor is reflected in agreements and the signing of a contract (Kamminga, 2008). Commitment has a formal and an informal side. The informal part consists of an implicit set of expectations between the client and the contractor. It is a highly flexible, continually changing and undefined set of terms that is called the informal psychological contract (Hoezen, 2012). The formal establishment and codification of commitment between client and contractor is formed by the formal legal contract. Both formal and informal agreements are part of the implicit and explicit knowledge base within a client organisation that is used to design the governance structure of new projects. These are accomplished through decision-making and sense-making processes during the procurement stage of a project.

Procurement processes

Procurement processes can be considered as decision-making processes in a particular legal setting (Volker, 2012). Procurement decisions are intentions for action that include an element of choice (Hodgkinson and Starbuck, 2008): for example, in tendering, a winner is chosen from all candidates that submitted proposals for a building project. Also, making prior announcements of decision criteria and decision methods could help to instantiate the basic EU principles of transparency, objectivity and equal treatment, and inform participating companies what to expect (Arrowsmith, 2005; Rowlinson and McDermott, 2005). However, procurement decisions also have to be considered as high-stake decisions due to their political sensitivity, the large sums of public money involved and the high impact of the built environment on the citizens' wellbeing (Cairns, 2008).

Procurement processes can also be considered as sense-making processes. In these, clients and contractors construct and reconstruct meaning through interactions with each other during the procurement stage, thus providing their understanding of the world and how to act collectively (Balogun *et al.*, 2008; Weick, 1995). People produce or reactivate accounts to deal with uncertainty and ambiguity, especially in dynamic and turbulent situations, and these are included in the mental models of individuals in order to make decisions (Basu and Palazzo, 2008; Maitlis and Sonenshein, 2010). Procurement decisions can thus be characterised as an interactive search, in which the representatives of a client body aim to find a contractor who can deliver the project that is desired. Essential to this process is the interaction between the decision makers and the stakeholders in order to make sense of the options (Kreiner, 2006) and to come to a mutual commitment. These processes therefore strongly contribute to learning processes in organisations.

Analytical framework

Procuring agencies could learn from (un)successful projects by studying the governance of these projects and by identifying the decisions in the procurement stage of the projects that were key to the governance of the projects. Since many researchers in project management literature display the difficulties of inter-project learning, we focus in this chapter on how procuring agencies can learn from sequential projects. We address the interrelationship of procurement processes and project governance, based on the assumption that commitment is created in a procurement situation and embedded in project governance structures within a client organisation. The empirical work is limited to public infrastructure projects for which procurement regulations apply.

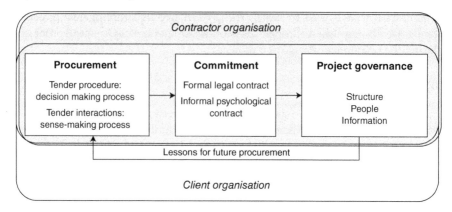

Figure 8.1 Client learning from project procurement and project governance through commitment

Our line of reasoning is built on the assumption that a procurement process (tender procedure and interactions) leads to a formal legal contract and an informal psychological contract between client and contractor. These commitments influence the manner in which the project will be governed. It is important to note that cooperative relationships are thus not only determined by the legal obligations, but also by the informal psychological agreements as understood at the signing of the contract and thus at the beginning of the project.

Figure 8.1 shows how we identify the procurement process as a major determinant for the commitment of the parties involved influencing the governance of a project. Organisational learning occurs by experiencing the successful or unsuccessful elements of existing governance mechanisms, which (ideally) then are incorporated in new tenders. Besides individual learning of project managers and tender consultants, who take on lessons learned in one project to their next project and adapt accordingly, institutions can learn by adjusting their policies and instruments. When the client–contractor relationship in one project is perceived to prevent a project from becoming a success, the client organisation will ideally think of manners for the governance of next projects to improve the project performance.

Research approach

Data collection and analysis

Based on the case of the Dutch Highways Agency, we describe how procuring agencies can take lessons from early projects on to future projects. Three complex infrastructure projects show how formal and informal aspects of commitment between project partners were established during the procurement phase, and how these influenced the client–contractor relationship within the three aspects of project governance: information, structure and people. The comparison of the cases indicates how the Dutch Highways Agency used the lessons of the projects when designing a new project.

Three major infrastructure projects (a tunnel and two highway routes) were used as rich empirical case studies involving particular but typical social situations developed from a variety of data sources (Easton, 2010). All three projects were initiated by the Dutch Highways Agency and can be considered as embedded cases (Yin, 2009). In each project the formal procurement requirements were similar. Nevertheless, formal and informal contractual processes differed. The cases were followed in time, enabling the client organisation to incorporate lessons learned in the procurement situation.

Most of the data was collected by interviews. In each project, the project leader and tender manager of the client and of the main contractor were interviewed. They all were involved in the procurement and the first stages of the case projects. On each project, two to four people from each party were interviewed, ranging from 60 to 90 minutes duration and based on semi-structured questioning. The semi-structured interviews involved general questions about the project characteristics, the legal contract details, the informal relational elements and the psychological development of the relationship. Interview data was captured by note-taking and digital recording that were transcribed verbatim in order to develop a comprehensive database of all three projects.

The data were initially analysed as separate case studies and then systematically compared across cases on constructs that emerged through the process as described by Yin (2009). Throughout data analysis and reporting the authors were frequently going back and forth between the interpretation and the original data. This process can be characterised as *ex ante* use of theory in qualitative research (Andersen and Kragh, 2010). The general aim of this approach is '*not to build consensus among diverging theoretical perspectives but rather to use their divergences as vantage points for creating new insights*' (Andersen and Kragh, 2010: 53). Therefore, we analysed the data to indicate which elements of commitment were key to the development of project governance structures based on the analytical framework as developed from theory (Figure 8.1). We first describe the character of the case and then discuss the findings according to our theoretical perspective on the learning curve between procurement, commitment, client–contractor relationship and project governance.

The case of the Dutch Highways Agency

As the executive organisation of the Ministry of Infrastructure and Environment, the Dutch Highways Agency maintains and develops national roads, waterways and open waters, and supports a sustainable environment. The organisation originated in 1798, when it was founded to take control of public works and water management. Nowadays, the organisation has around 8,800 employees, with an annual budget of EUR 5 billion. It manages 90,310 km² of surface water; 236 kilometres of dikes, dunes and dams; 5 storm surge barriers; 6,976 kilometres of canals, rivers and waterways on the open water; 3,076 kilometres of main highways, including traffic signalling systems; 2,843 viaducts; 24 tunnels; and 767 moveable and fixed bridges. The projects described in this case are construction projects meant to create additional highway capacity. The Dutch Highways Agency used several design–build–finance–maintenance contracts (DBFM) to govern these construction projects. In this kind of contract, the contractor is required to maintain the existing roads and tunnels, to design and build the additional capacity, and to find financial means to cover the costs. The Dutch Highways Agency pays a monthly fee for the availability of road capacity. After the building phase, when the road capacity is enlarged, a one-off payment is made that covers the construction costs. For the rest of the contract duration, the agency pays a monthly fee for the availability of road capacity and the contractor takes care of maintenance. All three projects included in this study were based on DBFM contracts.

The first project here is the Coen Tunnel project. The Coen Tunnel project consists of widening approximately 14 kilometres of highways at the north and south entrances to the existing 40-year-old Coen Tunnel, expanding the tunnel's capacity from two lanes to three in each direction plus two further reversible lanes, enabling five lanes of traffic in one direction during peak hours. The duration of the contract has been set at 30 years, from 2008 to 2036, with a contract value of EUR 600 million. The construction stage for the new tunnel started in 2009 and finished in 2014.

The second project is the extension of Highway A15, which connects Rotterdam Harbour to the European hinterland and is therefore an important traffic corridor in the Netherlands. Since a significant increase of traffic is expected due to the expansion of the harbour, the project consists of capacity extension of 37 kilometres of highway between de Maasvlakte and the Vaanplein crossing. The contract was tendered through a competitive dialogue just before the contract of the third case (Highway A12) was closed in December 2010. Total budget is around EUR 1,500 million, which makes it one of the biggest contracts ever awarded by the Dutch Highways Agency.

The third project is Highway A12. The Highway A12 project concerns part of the main traffic corridors of the Netherlands from the main ports (Schiphol Airport and Rotterdam Harbour) to the eastern part of the country, connecting Germany and other European countries. The project consists of reconstruction and capacity extension of the route Utrecht–Lunetten–Veenendaal in the middle of the corridor, 30 kilometres of highway in total. Part of the project belongs to measures taken by the Dutch government to boost the economy following the 2009 economic crisis. In the project the maintenance for the next 20 years will be executed by the same contracting consortium. The contract value is EUR 263 million.

Results

Procurement process

In formal terms, the procurement stage was identical in all three projects. The tender procedure was designed as follows: in three rounds of a competitive dialogue, parties were brought back from N candidates to three final bidders, of which one bidder was awarded the contract. Downsizing from N to 3 was based on dialogue products, handed in by the candidates and assessed by the procuring agency. These products included several aspects of the projects, such as schemes of action, risks inventories and management plans. The characteristics of the projects mainly differed due to the differences in project size, the time period and the project teams involved since the core elements of the procedure appeared to be satisfactory. The duration of the dialogue varied between 18 and 35 months, which induced high transaction costs for both the client and the participants.

In all projects the dialogue was aimed at project control, which reflected in the award criteria, starting with risks to be transferred from the agency to the future contractor. This resulted in many discussions on the budget conditions and limitations. In all projects, the award criteria turned out to be distinctive: parties came with varying views and solutions, both to the risk transfers and to the added award criteria. These views and solutions were based on the dialogues that happened to focus on the award criteria. In Project 1 dialogues focused on the case-specific risks; in Project 2 on technical solutions; and in Project 3 dialogues were focused on time gains. Time was not one of the award criteria. Yet, since the payment for availability of the road related to this issue, the candidates considered the bonus for time gains to be so high that their focus was on that aspect.

The *tender interactions* differed significantly, though. In Project 1, the award criteria raised questions of inconsistency and subjectivity, whereas participants in Project 2 and Project 3 were much more positive. This also had to do with the quality of the dialogues undertaken between the client and the future contractors. During the dialogues, contractors sought the agency's reasons for specific award criteria, came with possible solutions and asked for responses. It varied from project to project whether this worked or not. In Project 1 the agency's team were anxious not to make legal mistakes, so a lawyer attended all meetings and dialogues. In case of doubt, they would rather give no answer than risk misinforming the candidates. The agency's team took the rational approach to the tender process. They were careful not to develop a preference for one of the candidates and to be as objective as possible. Retrospective interviews indicated that the dialogue was characterised by participants as distant, formal and technocratic, which did not help them to come to an understanding of the agency's actual goals and preferences for any solution beforehand. It also contributed little to making sense of the project. This was reflected in the formal questions asked: 23 per cent of the questions regarded the procedure itself. In Projects 2 and 3, this number of questions on the procedure decreased to 12 per cent and 8 per cent respectively.

Candidates were much more positive about the dialogues in Projects 2 and 3. In Project 2, there was more attention on soft aspects of cooperation, reflected in the award criterion 'cooperation and distribution of roles'. Furthermore in Projects 2 and 3, the agency was already more experienced with the competitive dialogue, and better aware of the legal consequences of its actions. The teams that were involved in the dialogue felt more confident with regard to the restrictions of objectivity and equal treatment. They were therefore able to operate in a less formal and distant manner than the team in Project 1. The candidates in Projects 2 and 3 reported in the retrospective interviews that the dialogue fundamentally contributed to their ability to make sense of the project and to determine the aspects that mattered most to the agency.

Commitment

Regarding the *formal legal contract*, we found that the incentives in all three projects were to deliver the construction works as soon as possible: the contractor would receive payment on the date that the new constructions would open. Furthermore, the roads needed to be accessible at all times. Contractors received payment for accessibility; fines for non-accessibility were set within the contract. However, experiences now show that this kind of incentive mechanism has limitations. Early delivery before the estimated date could, for example, seriously change the project structure, which would increase the risks of other elements in the project due to interdependencies in the construction process, such as between permits and groundwork. Since the limits of acceleration were mainly negotiated out during the procurement process, the latitude for additional activities appeared to be narrow.

Following on from Project 1, it was decided that Project 2 would have one additional bullet payment moment due to the change in economic conditions. This fitted the scale and character of the project and increased the options of the consortia financing the project. In Project 3 the incentive schemes that related to the delivery dates were used by the consortium to innovate the engineering and construction process of overpasses. The additional costs for the development of this new construction method balanced the additional revenues, but reduced the risks of late delivery.

The contract clauses used in Project 1 were less mature than the clauses in Projects 2 and 3, which showed the learning curve of the awarding authority. In terms of opportunity control, project 1 constrained the opportunities for contractor's strategic behaviour the least, while project 3 controlled the opportunities the most. This was mostly due to the output specifications of the contracts. The specifications in Project 1 were not complete because of concern with the technical demands. Further, the process specifications were non-specific and open to interpretation. The contract for Project 2 was more SMART (meaning specific, measurable, achievable, relevant and timed) on the technical specifications, and it contained added system demands. The process specifications became even more SMART with each project where they were applied. Especially, the specifications for Project 3 were extended with lean and mean system demands and specific requirements for a dynamic traffic management system. Yet, searching for a concrete performance measurement system remains a challenge, especially in the realisation of the agreements. With regard to the monitoring system, the three contracts are comparable. All contracts made use of a monitoring system based on the project management system (PMS) of the contractor. As part of a risk assessment, the client checked the contractor's system, its processes, or specific products.

In contractual terms several standards were agreed. However, our results indicate that building relationships are of critical importance because the formal agreed arrangements and systems do not completely fit reality. This confirms the presence of sense-making processes, which change the mental models of the actors involved in the projects. In Project 3, both parties were flexible in finding a solution for the problems that arose during the project, while in Projects 1 and 2 these discussions led to considerable differences of opinion. The tone of such discussions appears to have been set during the procurement phase and stemmed from problems of understanding and information asymmetry. For the exchange of information, the agency relied on several management systems. In all projects, these systems provided the information that was input for communication about the project progress. In general, this worked fine for them. However, Project 2 showed that, when there is lack of understanding about what both parties can expect from the management system, it is not only a source of information, but also a possible cause for disputes.

Regarding the *informal psychological contract,* the procurement stage of the three projects differed significantly, leading to various levels of benevolence at the start of the projects. Whereas the contractors of Projects 2 and 3 shared mutual understandings with the client, this was not the case in Project 1. Here, contractual renegotiations, disputes and legal processes occurred even before the contract was signed resulting in a lack of benevolence. By the time the contract was signed, it was clear to both parties that their ideas of the project content and norms concerning their relationship differed. This resulted in a complete lack of empathy and affection between the parties. The complete opposite appeared in Project 3. Here, both parties invested in growing mutual understandings during the procurement stage, resulting in shared norms and values, and creating a basis for empathy between client and contractor, which supported the constructive relationship during execution of the contract. In Project 2, the level of shared understandings at the time the contract was signed was average. Norms and values were comparable, and there was a kind of distant empathy between the contractor and its client. However, recent experiences have shown that problems of understanding occurred around the contractor's quality measurement system (and, with that, conflicts about payment), thus putting pressure on the relationship. After almost a year of conflicts, both the contractor and the agency replaced some of their key personnel. After this, the agency worked alongside the contractor in a cooperative yet strict manner (government payments need to be accounted for very carefully), so that the project could be managed more easily with distant cooperation.

Although all three projects were contracted to a one-off consortium involving more than one company, the non-material incentives to utilise opportunities for opportunism differed between the parties. Project 3 was contracted to a consortium that consisted of cooperating subsidiaries of one large Dutch construction firm, and thus was always associated with the parent company. The on-going relationship and reputation were much more important to this contractor than to contractors in the other two projects, where the contractors were consortia existing of several cooperating firms. Nevertheless, the companies involved in the consortium of Project 2 regularly worked together, and the companies in Project 1 were also active in other infrastructure projects, either singlehandedly or in consortia.

All three consortia felt that their project performance would affect the attitude of the client towards the cooperating companies in future large projects. The contractor in Project 3 thought it would suffer from a bad reputation in both large and small future projects, involving both the parent company and its subsidiaries. We learned, however, that the importance of reputation management should not be overestimated. Despite the fact that a need exists to take the shadow of the future and the shadow of the past into account when selecting contracting parties, this has not been officially implemented yet in the Netherlands. The main reasons for this are the limited room for discrimination in current European procurement regulations and the complexity of measuring performances and comparing results among projects (Petit, 2010). Preliminary experiences in other Dutch client organisations do show that the differences between contractors are generally minimal, so the discriminating value of such instruments appears to be limited.

Furthermore, we found that routines (informal understandings about the working manners in the specific projects) differed between the client–contractor teams in all three projects. In Project 3 the contractor invested in the internal routines, which in turn led to the quick development of routines in the relationship with the client. These routines were little developed in Project 2, and even less in Project 1. The results indicate that this was caused by little previous co-working experience between the members of the consortium. Neither the contracting firms nor the client had invested in developing routines. Even though in Project 2 specific attention was given to cooperation and role distribution, the difference in interests turned out to be major, overshadowing the cooperative intentions of the parties involved. We believe that the more formal incentive control mechanisms are internalised, the less there is a need to make use of them during conflicts and in the final settlements.

Project governance

Regarding the *structure* of the projects, all three projects involved a DBFM contract, which requires the engineering, construction and maintenance for a considerable amount of time. In each project the client and the consortium assigned a project leader and a contract manager, who communicated about the issues arising during the project. The client was not concerned with the project structure as long as the consortium delivered according to plan. In Project 3, the financial institution required the consortium to govern the project as one integrated engineering and maintenance organisation during construction. During maintenance, the engineering project company would be taken out of the consortium. They could fulfil this need relatively easily since most partners belonged to the same corporation. This was not the case in Projects 1 and 2. These consortia were built from different companies with different backgrounds and organisational cultures. This appears to have caused more friction within the consortia and seemingly also more conflicting relations between the client and the consortium.

The strict financial plan of a DBFM contract compels the consortium to think about maintenance at an early stage of the project. In all projects, deviations in functionalities were usually discussed with the client first and then financially calculated in a business case for the bank. According to the project leader of Project 3 this even *'makes it fun to think about maintenance'*, which shows a shift in the perception of the underlying importance of new construction works in relation to the integrity of existing works. Yet, in general, the structure of these projects did not show very much integration of building phases. So, despite the theoretical added value of integrated contract and continuous discussions on costs from a life cycle perspective during the procurement phase, results indicate that only during the execution of the project does awareness arise about the value of integration.

Results indicate that people make a large difference in project governance. In Project 1, both client and contractor teams had negative attitudes to each other based on a conflict that occurred in the time between the final bid and the contract award (for a detailed description, see Hoezen, 2012). The cooperation started somewhat hostilely, and it took a while before the air cleared. It helped that the agency's project team was changed during the project. The main reason for this change was that the knowledge of the first members of the client's project team was only needed for the initialisation of new DBFM projects. This concerned not only projects prepared by the agency itself. First, a large number of people were hired as consultants and so they started searching for new projects after the contract was awarded. Second, on the contractor's side, the project members' experience was needed in other projects. Besides, the actual start of the work asked for different specialisations in each stage of the project. The contractor in Project 1 organised the work in a rather traditional manner, operating as one main contractor, and dividing the work into work packages for several sub-contractors. It took them some time to find a modus operandi in their relationship with the stakeholders, which was a cause for changes in the general project team as well.

Despite the innovative procurement and contracting relations, Project 2 can also be considered as a rather traditional project organisation. The project leaders and contract managers from both sides regularly communicated about evolving issues, keeping in mind which issues were the responsibility of whom. The problems that arose in the construction phase indicate a delicate relation between both parties involved in the contract. Based on the current data it is hard to elaborate on the actual reasons for this, but we assume that communication during the procurement stage did not focus on how the contractor and the client understood their formal agreement. Furthermore, the attitude of the project leaders appears to be an important factor. Although cooperation and distribution of roles were award criteria during the tender, the consortium that was awarded the contract for Project 2 merely won the tender due to their low bid.

In the construction stage of Project 3 both sides of the project team (client and constructor) started with getting to know each other 'to make them trust you'. Consequently an open, constructive and cooperative environment was built. A statement by the project leader of the consortium – '*I don't have secrets, they are allowed to see everything*' – shows the level of trust and openness of the project leader. It is also in line with the course that the Dutch Highways Agency has set for the future in collaborating with suppliers. Results indicate that the key players in this project were selected based on their cooperative attitude. According to the project leaders from both the contractor and the client this trust-based attitude in combination with 'I enjoy this profession' also reflects upon the rest of the project members, monthly some 3–500 people, leading towards a unique cooperative project climate that has been awarded several awards lately.

In all projects information was transferred via specific management systems that were agreed upon during the tender. Since all three projects involved replacement or maintenance on existing assets as well as construction of new infrastructure, considerable risks needed to be taken by the consortia. The data indicate that openness about these risks requires a trustful environment. Given the high level of legal awareness in Project 1, the agency assessed each request for information during the tender on the possibility that a party could use the information as an option for claims during the realisation stage of the project. Due to this, the risk provisions were high and the solutions less optimal than possible. In Project 2 it was found that, when contractor and client shared information on payment risks, the solution to the conflict over the quality management system became closer. This shows that, although in case of (possible) conflicts it seems strategically right not to share information, the solution to problems of understanding does lie in sharing. In Project 3 this worked out very well: the investment in creating shared understanding by sharing information led to cooperation between the agency and the contractor. Both parties relied on the management systems to their mutual satisfaction.

Conclusion

Based on three complex infrastructure projects, we explored how learning experiences from procurement processes influence project governance within one client organisation. The differences found in the projects related mainly to problems of understanding such as contract clauses, grounds for fines (Project 1) and specifications of the quality management system (Project 2). In response to conflicts, the teams renegotiated the formal agreements (Projects 1 and 2), replaced key personnel (Project 2) and started conversations to overcome information asymmetry (Project 2). Due to the openness within the client organisation on these matters, the project teams in Project 3 were able to anticipate these possible differences in understanding by investing in a cooperative relationship from the beginning of the project. Although these problems of understanding cannot be fully avoided, results indicate that client organisations can become more sensitive in providing room for prevention and act upon their experiences in situations that would otherwise escalate.

Concerning the commitment and relationship building, we saw that a shared organisation culture can stimulate the development of internal routines. This is strengthened by the explicit vision to steer not only on formal structures, but also on people and (trust in) a reliable information system. Without room for sense-making in the realisation of these agreements, sincere tensions develop between both parties. Searching for a win–win situation in an open project climate proved to be most beneficial for gaining commitment as can be seen from the experiences of Project 3. This attitude requires not only formal approvals from senior management from the beginning but also perseverance of all employees and willingness to change the organisational culture, which cannot be found in every organisation, especially not in a traditionally oriented sector such as construction.

We found that in establishing the project governance structure both parties can set the tone in the 'design' of the project governance, and provide the people and information systems that support this structure. Each project required specific adjustments for new approaches and governance structures. Yet, learning from previous projects increases the level of sense-making and understanding, which then again enhances the possibilities to 'master' the design process by reducing uncertainties of the actors involved.

A formal evaluation of each project on the procurement and project governance aspects of projects can contribute significantly to the learning capacity of a client organisation. This would also open up interesting opportunities for further research, since we only included three projects of one client in this study. Information for these kinds of evaluation does not have to be collected separately. For example, Verweij *et al.* (2015) recently explored correlations within the existing project databases of the Dutch Highways Agency. If other clients would also open up their databases, it would be possible to make comparisons between clients in different sectors and different countries. This would expand the learning capacity *within* a client organisation to learning *between* client organisations, contributing to increased performances in the construction industry in general. Furthermore, it would be interesting to see not just how the client organisation learns and adapts after a first project, but also how the contracting organisations learn and anticipate on their project experiences.

References

Andersen, P. H. and Kragh, H. (2010). Sense and sensibility: Two approaches for using existing theory in theory-building qualitative research. *Industrial Marketing Management*, Vol. 39 (1), 49–55.

Arrowsmith, S. (2005). *The Law of Public and Utilities Procurement* (2nd edn). London: Sweet & Maxwell.

Atkin, B., Flanagan, R., Marsh, A. and Agapiou, A. (1995). *Improving Value for Money in Construction: Guidance for Chartered Surveyors and their Clients*. London: Royal Institution of Chartered Surveyors.

Balogun, J., Pye, A. and Hodgkinson, G. P. (2008). Cognitively skilled organizational decision making: Making sense of deciding. In: Hodgkinson, G. and Starbuck, W. H. (Eds.) (2008). *The Oxford Handbook of Organizational Decision Making*. New York: Oxford University Press, pp. 233–49

Basu, K. and Palazzo, G. (2008). Corporate social responsibility: A process model of sensemaking. *Academy of Management Review*, Vol. 33 (1), 122–36.

Bosch-Sijtsema, P. M. and Postma, T. J. B. M. (2010). Governance factors enabling knowledge transfer in interorganisational development projects. *Technology Analysis & Strategic Management*, Vol. 22 (5), 593–608.

Cairns, G. (2008). Advocating an ambivalent approach to theorizing the built environment. *Building Research & Information*, Vol. 36 (3), 280–89.

Caldwell, N. D., Roehrich, J. K. and Davies, A. C. (2009). Procuring complex performance in construction: London Heathrow Terminal 5 and a Private Finance Initiative hospital. *Journal of Purchasing and Supply Management*, Vol. 15 (3), 178–86.

Dorée, A. G. (2001). *Dobberen tussen concurrentie en co-development: de problematiek van samenwerking in de bouw (in Dutch: Floating Between Competition and Co-Development, Inaugural Lecture)*. Enschede: University of Twente.

Easton, G. (2010). Critical realism in case study research. *Industrial Marketing Management*, Vol. 39 (1), 118–28.

Eriksson, P. E. and Westerberg, M. (2011). Effects of cooperative procurement procedures on construction project performance: A conceptual framework. *International Journal of Project Management*, Vol. 29 (2), 197–208.

Hartmann, A., Reymen, I. M. M. J. and Van Oosterom, G. (2008). Factors constituting the innovation adoption environment of public clients. *Building Research & Information*, Vol. 36 (5), 436–49.

Haugbølle, K. and Boyd, D. (2013). *Research Roadmap Report: Clients and Users in Construction*. CIB Publication 371. Rotterdam: CIB.

Hellström, M., Ruuska, I., Wikström, K. and Jåfs, D. (2013). Project governance and path creation in the early stages of Finnish nuclear power projects. *International Journal of Project Management*, Vol. 31 (5), 712–23.

Hobday, M. (2000). The project-based organisation: An ideal form for managing complex products and systems? *Research Policy*, Vol. 29 (7–8), 871–93.

Hodgkinson, G. and Starbuck, W. H. (Eds.) (2008). *The Oxford Handbook of Organizational Decision Making*. Oxford: Oxford University Press.

Hoezen, M. (2012). *The Competitive Dialogue Procedure: Negotiations and Commitments in Inter-organisational Construction Projects*. PhD Dissertation. Enschede: University of Twente.

Hoezen, M. and Volker, L. (2015). The importance of procurement negotiations for project success. *IPMA Projectie Magazine*, Vol. 1, 2–8.

Kamminga, Y. P. (2008). *Towards effective Governance Structures for Contractual Relations; Recommendations from Social Psychology, Economics and Law for Improving Project Performance in Infrastructure Projects*. PhD Dissertation. Tilburg: Tilburg University.

Kreiner, K. (2006). *Architectural Competitions: A case-study*. Copenhagen: Center for Management Studies of the Building Process.

Lizarralde, G., Tomiyoshi, S., Bourgault, M., Malo, J. and Cardosi, G. (2013). Understanding differences in construction project governance between developed and developing countries. *Construction Management and Economics*, Vol. 31 (7), 711–30.

McWilliams, A. and Smart, D. L. (1993). Efficiency v. structure-conduct-performance: Implications for strategy research and practice. *Journal of Management*, Vol. 19 (1), 63–78.

Maitlis, S. and Sonenshein, S. (2010). Sensemaking in crisis and change: Inspiration and insights from Weick (1988). *Journal of Management Studies*, Vol. 47 (3), 551–80.

Mason, E. (1939). Price and production policies of large-scale enterprise. *American Economic Review*, Vol. 29 (1), 61–74.

Petit, C. H. N. M. (2010). Past and Present Performance. *Tijdschrift voor Bouwrecht*, 8 (143), 772–83.

Ring, P. S. and van de Ven, A. H. (1994). Developmental processes of cooperative interorganizational relationships. *The Academy of Management Review*, Vol. 19 (1), 90–118.

Rowlinson, S. and McDermott, P. (Eds.) (2005). *Procurement Systems: A Guide to Best Practice in Construction*. London and New York: Routledge.

Ruuska, I., Ahola, T., Artto, K., Locatelli, G. and Mancini, M. (2011). A new governance approach for multi-firm projects: Lessons from Olkiluoto 3 and Flamanville 3 nuclear power plant projects. *International Journal of Project Management*, Vol. 29 (6), 647–60.

Tabish, S. Z. S. and Jha, K. N. (2011). Identification and evaluation of success factors for public construction projects. *Construction Management and Economics*, Vol. 29 (8), 809–23.

Thiry, M. and Deguire, M. (2007). Recent developments in project-based organisations. *International Journal of Project Management*, Vol. 25 (7), 649–58.

Verweij, S., van Meerkerk, I. and Korthagen, I. A. (2015). Reasons for contract changes in implementing Dutch transportation infrastructure projects: An empirical exploration. *Transport Policy*, Vol. 37 (0), 195–202.

Volker, L. (2012). Procuring architectural services: Sensemaking in a legal context. *Construction Management and Economics*, Vol. 30 (9), 749–59.

Weick, K. E. (1995). *Sensemaking in Organizations*. Thousand Oaks, CA: Sage Publications.

Yin, R. K. (2009). *Case Study Research: Design and Methods* (4th edn, Vol. 5). Beverly Hills, CA: Sage Publications.

9 Quality and satisfaction with constructed roads in Nigeria

The clients' view

Chimene Obunwo, Ezekiel Chinyio,
Subashini Suresh and Solomon Adjei

Introduction

Construction clients are individuals or groups of people who appoint and assign others to carry out construction projects. The CIB W118 research roadmap identifies clients as operating within a wider environment of businesses and society, ensuring that construction needs are provided. This roadmap outlines the broad spectrum of construction activities, stretching its focus from demand and supply to identifying clients' needs (Haugbølle and Boyd, 2013). As stated by Vennstrom and Eriksson (2010) it is the duty and responsibility of the construction client to see that the users' needs or expectations are turned into physical reality.

The terms customers and users are often used interchangeably to represent the group of people who either pay for or use a product or service (Torbica and Stroh, 2000). In this chapter 'users' denote the people who benefit from the use of a constructed project. Users may not necessarily pay directly for the constructed product. When considering public construction projects, the government is usually the main client that employs the services of contractors to deliver projects for use by the public. The public latently pay for construction projects especially through taxes or other means of internally generated revenue, and their satisfaction is thus vital to the success of projects (Adenugba and Ogechi, 2013). Meanwhile, the quality of constructed roads has been identified as a key determinant of project success. It is thus necessary to make sure that users' needs are incorporated in road construction processes in order to deliver high quality projects that reflect their desires. This is one of the primary responsibilities of the construction client, which makes it incumbent on the client to identify all areas of constructed roads that are relevant for client satisfaction.

From a global perspective, the need for satisfaction has been identified as a key issue with construction stakeholders. Indeed, current research trends suggest that all relevant factors should be explored and considered to ensure the delivery of high quality construction projects that satisfy the needs of the concerned stakeholders (Cheng *et al.*, 2005; Idoro, 2010; Rahman and Alzubi, 2015). Traditionally, these stakeholders include architects, project managers, civil engineers, government regulatory bodies, clients and users, amongst others (Al Nahyan *et al.*, 2012; Obunwo *et al.*, 2013). Central to the stakeholder group is the client who categorically owns the project and assigns construction activities mainly to contractors.

This chapter presents findings from a research project which studied the extent to which a number of attributes of project quality could predict and consequently enhance client satisfaction within road construction projects in Rivers State, Nigeria. The research specifically aimed at finding out which attribute of project quality had the greatest impact on the level of client satisfaction. Three attributes of project quality, namely project performance, project reliability and project aesthetics, were studied and regressed with two attributes of satisfaction, which were contractor referral and contractor repatronage (Idoro, 2010; Masrom *et al.*, 2013; Xiong *et al.*, 2014). The percentage contributions of these attributes of project quality that predict or enhance client satisfaction are therefore presented to inform construction stakeholders on areas to relatively channel their efforts while developing and managing new road construction projects, especially in Rivers State, Nigeria. This chapter begins by looking at the Nigerian construction industry in context to determine the opportunities and challenges therein. The attributes of project quality are then discussed, followed by measures of client satisfaction. The methodology adopted in the research is then presented prior to the findings of the study. The chapter ends with a discussion of the findings and highlights some implications for future practice.

The Nigerian context

Nigeria is the sixth largest exporter of crude oil (OPEC, 2010) and uses its petro-dollar income to undertake construction and other activities. The Nigerian construction industry alone accounts for 1.4 per cent of its GDP (6.6 per cent in 2011) (NPC, 2012). Sanni and Windapo (2008) highlight that the Nigerian construction industry occupies a significant portion of the capital base of the national economy, adding that its success or failure has huge positive or negative impacts respectively. The Nigerian construction industry basically consists of two main sectors:

- a formal or organised sector, made up of organisations legally registered in the country and operating under set laws on employment, procurement and tendering, combining both highly skilled and expatriate labourers in their operations; and
- an informal or unorganised sector, made up of owner-supervised constructors of simple residential buildings and similar structures; operating by using private citizens and gangs of artisans (Aniekwu, 1995).

Despite the gigantic infrastructure in the country, the Nigerian construction industry is declining (Oluwakiyesi, 2011; Odediran *et al.*, 2012). This notwithstanding, participants in the Nigerian construction industry have in recent years progressed from assessing project delivery via the 'iron triangle' of cost, quality and time, to incorporating satisfaction in the project success criteria

(Ogunlana, 2010). This aligns with the expectations of the CIB W118 research roadmap. Both the clients who procure the construction activities and the users of the eventual products need to be satisfied with the constructed facility in order to optimise the success of the associated construction project (Boyd and Chinyio, 2006). Despite the increasing need to optimise the expenditure on resources while executing road or other construction projects, the issue of satisfaction has continued to gain research attention (Karna, 2014; Rahman and Alzubi, 2015).

The UN-Habitat (2014: 73) has indicated that '*almost 60 % of the urban population in sub-Saharan Africa resides in slums without running water, minimum conditions of hygiene and without access to essential services*' such as roads; also that, by 2025, 60 per cent of the population of less developed countries would be living in urban areas (UN, 2014). Obinna *et al.* (2010) identify that there has been a reduction in the life span of urban roads and attributes this to the increase in the volume of road users due to the increasing migration of individuals to cities and urban areas. Within the Port Harcourt metropolis in Nigeria, for instance, there has been an increase in the commuting needs of its growing population (Obinna *et al.*, 2010). Port Harcourt is the capital of Rivers State, which is one of the 36 states in Nigeria. Port Harcourt hosts diverse oil companies and construction firms and hence a desired destination for many individuals and firms seeking to benefit from its huge potential economic gains.

Over the past two decades, there has been a 50 per cent growth in the population of Port Harcourt, where road transportation has become the most common means of commuting (Olatunji and Diugwu, 2013). Although road construction firms within the area constantly update their resources and activities to ensure satisfaction from the constructed roads, the huge traffic of individuals and machinery and the increasing commuting needs indicate that road construction projects within Port Harcourt are seemingly a never ending process. This is primarily as the government, the major client and other construction stakeholders strive to continuously meet the needs of the public and guarantee satisfaction with the constructed roads.

Opportunities and challenges facing the Nigerian construction industry

The opportunities within the Nigerian construction industry are anchored on the impact of construction on economic development (Oluwakiyesi, 2011). Construction activities are achieved through huge supply chains that utilise various skilled and unskilled labour, materials, equipment and machinery as well as financial transactions that enhance the Nigerian economy. The government, which is the major client for construction activities in Nigeria, thus continues to initiate road and other types of project to keep the economic machinery going. When compared to the UK and Chinese construction industries, the economic opportunities from construction are evident. The UK Contractors Group (UKCG), which is the primary association for contractors and their supply chains, has highlighted the importance of construction to economic growth: thus, every GBP 1.0 spent on construction generates a total of GBP 2.84 in total economic activity, resulting in an increase in the GDP (UKCG, 2012). This UKCG report of 2012 identified construction as a major contributor to the UK GDP (estimated at 7.4 per cent in 2011), while Rhodes (2015) reports that 6.5 per cent of the total UK economy in 2015 was attributed to the UK construction industry's output, amounting to GBP 103 billion. Similarly, 13 per cent of the Chinese GDP is attributed to the construction industry (Oluwakiyesi, 2011). China has recorded an impressive growth in construction as a percentage of the GDP from 3.8 per cent to 6.6 per cent, further showing the economic opportunities in their construction industry.

In addition to the economic opportunities, the reliability of road construction projects, which is the expected life span prior to failure (Masrom *et al.*, 2013), can be optimised through improvements in the construction activities in Nigeria. Construction stakeholders are therefore driven to develop strategies that identify critical areas where resources could be channelled to enhance the quality of constructed roads, benefit from its economic opportunities as well as improve on the nature of satisfaction recorded by its customers.

The implications for satisfaction within the Nigerian construction industry arise from the current state of the industry. This includes the economic impact of construction satisfaction as well as the impact of governance regarding construction. Despite the growth seen in the Nigerian construction industry, its contribution to GDP has remained at abysmally low levels in the past three decades due to barriers, such as corruption, shortage of technical expertise and general laxity, that hinder the progress and success of projects (Oluwakiyesi, 2011; Obunwo *et al.*, 2013). Odediran *et al.* (2012) explain that the Nigerian construction industry with projects that provide basic amenities such as shelter, roads, water and electricity is still in its infancy in terms of development, where the government is the major client. Odediran *et al.* (2012) further decry the poor growth of indigenous construction firms in Nigeria due to public-sector clients' preference for using international firms, although this is also responsible for the impressive growth of the Nigerian construction industry through enhanced government spending and provision of infrastructure throughout the country.

Despite these challenges, as well as the growing transportation and commuting needs, Nigerian construction firms seem to be misinformed on areas that need extra attention especially while delivering road projects generally and in Port Harcourt specifically. This informed the need to study areas of construction management that need stronger attention especially by identifying the contributions and predictive capability of the attributes of project quality towards client satisfaction.

Attributes of project quality

Several attributes are often identified to describe project quality in construction. However, two main and dissimilar features that have been considered in this regard are: quality as conformance to set out requirements, and quality as a prerequisite for client satisfaction. In other words, to achieve project quality and ensure satisfaction, there has to be a synergy between conformances to any benchmarked specifications and meeting the requirements of the client (Karna, 2014). The benchmarked specifications are usually set out by the designer, architect, contractor or a specialist in the project team, and achieving this sets the pace for client satisfaction. Within the scope of this chapter, the attributes of quality are summarised under three sub-headings:

- construction project performance;
- reliability; and
- aesthetics.

Construction project performance

Construction project performance entails the activities which ensure that the project fulfils its intended purpose and satisfies the expected needs (Oluwakiyesi, 2011; Olatunji and Diugwu, 2013). Atkinson *et al.*, (1997) and Love (2002) elucidate that successful construction project performance is achieved when the requirements of stakeholders are met individually and collectively. The measurement of performance therefore gives an indication of the success or failure of a project, although the identification of performance indicators, performance measures and the actual performance measurement are an arduous task (Beatham *et al.*, 2004). Items such as quality, functionality and the satisfaction of end-users, clients and design and construction teams are subjective measures of performance. On the other hand, objective measures of performance include construction time, unit costs and net present value. The broad constructs used to measure performance would include concerns on project cost, organisational team work, project design specifications, project abandonment, patronage from suppliers, health and safety considerations, incorporating the commuting needs of customers into construction, and enhanced project supervision, as well as the nature of conflict resolution in terms of handling complaints. The knowledge and measurement of project performance aids in evaluating construction activities and further enhances continuous improvements.

Construction project reliability

In any construction endeavour, the reliability of the project is a very essential and strong determinant of satisfaction. Reliability is defined as the probability that an item would perform to its required function without failure and within the constraints of specifications and time (Oluwakiyesi, 2011; Olatunji and Diugwu, 2013). Reliability can also be seen as the specified timeframe in which a system would consistently perform its intended function under specified environmental conditions without degradation or failure. Statistically, reliability is expressed as the mean time between failures (Jonsson and Svingby, 2007). Although Oluwakiyesi (2011) and Olatunji and Diugwu (2013) agree that structures and in this case construction projects may fail at any time, the nature of the design, its construction and mode of operation usually determine how reliable such a project would be. Despite the advancements in construction management, authors such as Gwilliam *et al.* (2009), Oluwakiyesi (2011) and Olatunji and Diugwu (2013) argue that the quality and reliability of construction projects are becoming more significant and these are seen as motivating factors for clients to either engage or withdraw from construction. In summary, ascertaining reliability in projects would involve the use of professional experts in the areas of design, monitoring, maintenance, inspection and benchmarking (Ali *et al.*, 2013).

Construction project aesthetics

Project aesthetics is a vital attribute of quality and it entails the physical and visible aspect of the constructed project (Pheng and Chuan, 2006). The aesthetic component of a construction project is also an indication of its architectural and design quality as well as the construction processes employed (Pheng and Chuan, 2006). Considering road projects, although aimed at providing structural integrity, the design is also expected to include artistic and appealing components as these are what would be visible to both the client and users. In the course of delivering construction projects therefore, contractors should be very particular about the aesthetics of the product because it offers a basis of assessing the quality of the product. It is, however, a common mistake by clients to misconstrue the aesthetic content of a construction project for integrity, lustre or expensive cost as aesthetics is a function of the materials used, the finishing employed and the adherence to design specifications. In summary, project aesthetics subsumes the features of implementing quality standards and technological advancements; considerations of using efficient materials, producing durable finishing and health and safety; as well as post project evaluations.

Measures of client satisfaction

Public construction projects involve activities proposed by a client but carried out by contractors that are carefully selected before a contract is awarded. It is the duty of the contractor to meet the demands of the client both in delivering the designed project and in obtaining satisfaction from the constructed project. Consequently, despite using tendering processes and complying with regulations surrounding contract award, contractors who have met or exceeded construction expectations in previous projects often place themselves in an advantageous position to be given another opportunity to perform. Hence the general perception is that, when construction stakeholders are happy with certain contractors' performance, they may either repatronise or refer them to other clients.

Contractor repatronage refers to a situation where a contractor is called upon to carry out a construction project based on the feedback from previous construction activities (Weil, 2005; Whittaker *et al.*, 2007; Pritchard *et al.*, 2009; Eadie and Graham, 2014). On the other hand, referral is when a contractor is given a reference by someone else during tendering in a bid to influence their chance of success (Idoro, 2010; Masrom *et al.*, 2013; and Xiong *et al.*, 2014).

Satisfaction from a constructed project can thus be estimated from the clients' rating of the contractors they have used and willingness to provide references for them. Every project is unique in terms of the combinations of risks and challenges (Koster, 2010). Consequently, a situation whereby a construction contractor is given referral or repatronised by a client based on previous work carried out would signify some high level of satisfaction obtained from their constructed projects. The possible linking of referral and repatronage to client satisfaction was studied empirically.

Research methodology

In order to obtain data for the research, a questionnaire was developed based on the attributes of:

- project quality, i.e. performance, reliability and aesthetics; and
- client satisfaction, i.e. contractor referral and repatronage.

These attributes were identified from an extensive review of literature and used as the basis for determining client satisfaction as they serve as a direct indication of quality with regards to the finished project (product).

Respondents were asked to rate their perceptions of these attributes on a 5-point Likert scale which ranged from 1 for strongly disagree to 5 for strongly agree; and 1 for highly dissatisfied to 5 for highly satisfied (Jonsson and Svingby, 2007). This strategy was adopted to obtain ordinal data as well as a true and realistic view of the respondents' preferences to satisfaction in road construction projects.

A pilot study was carried out to ascertain the relevance of the questions to the research aim and this led to the refinement of the research instrument. A total of 600 questionnaires were distributed to stakeholders of road construction projects within Port Harcourt: 518 were returned, of which 503 were found useful, a response rate of 83 per cent. This impressive response rate was due to distribution of the questionnaires in person to each of the respondents. The organisations surveyed were identified from the list of registered road construction companies in the Rivers State ministry of works, while a demographic assessment of the respondents aided in ensuring that the right people were surveyed while carrying out the research.

A comprehensive presentation of the data obtained in the course of the research is presented in Obunwo (2016). However, only data from the 116 respondents who were clients were used for the analysis presented here.

To evaluate the relationship between the dependent and independent variables and hence the contribution of each variable of project quality to client satisfaction, Step-wise Multiple Regression Analysis (SMRA) was employed while the statistical tool of SPSS was used to ease data presentation and computation.

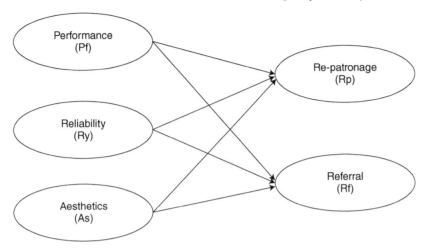

Figure 9.1 A 3×2 matrix for determining relationships of satisfaction

Soetanto and Proverbs (2001) argue that SMRA is a valuable means of predicting satisfaction in construction research and can be employed when the need arises to predict the value of a variable based on two or more other variables. The coefficient of determination (R^2), a statistical measure which indicates the relevance of the regression line in the approximation of real data points, was thus calculated.

The relationship between the attributes of project quality, which determines how a project performs in terms of the expectation of clients, and attributes of client satisfaction was expected to provide a good indication of the level of satisfaction. A 3×2 interaction of relationships was explored in the analysis (Figure 9.1). Firstly, the strengths of relationships between repatronage and the trio of performance, reliability and aesthetics were sought. After that the strengths of relationships between the trio and referral were analysed. While the trio of performance, reliability and aesthetics represented project quality, the duo of repatronage and referral represented the dimension of customer satisfaction.

Findings

The perceptions of the clients regarding their satisfaction are presented here. Influences on repatronage and referral are used accordingly.

Project aesthetics as the predominant predictor of contractor repatronage

Table 9.1 provides results from the first analysis which identified project aesthetics as the predominant predictor of contractor repatronage. Aesthetics alone accounted for approximately 46 per cent of the variance in repatronage. Further analysis revealed that the beta values for performance, reliability and aesthetics were highly significant. The associated equations indicated that any increase in

Table 9.1 Model summary of stepwise regression analysis on the relative contribution of project quality variables to customer satisfaction (repatronage)

Model	R	R Square	Adjusted R Square	Std. Error of the Estimate
1	0.682[a]	0.465	0.460	9.41752
2	0.695[b]	0.483	0.474	9.25273
3	0.706[c]	0.498	0.484	9.15665

Notes:
a. Predictors: (Constant) Aesthetics
b. Predictors: (Constant) Aesthetics, Reliability
c. Predictors: (Constant) Performance, Reliability, Aesthetics

the value of any of the independent variables will yield a resultant increase in the value of repatronage.

Project aesthetics as a significant predictor of contractor referral

The next analysis explored the strengths of relationships between project quality (performance, reliability and aesthetics) and the other element of client satisfaction (referral). Although a tendering process is often used in construction contracting, referrals can augment the identification and/or shortlisting of competent contractors. In this second analysis, aesthetics alone accounted for more than half of the variability, approximately 65 per cent of the variance in contractor referral, while project performance and reliability were weaker predictors.

Table 9.2 shows the results of a further analysis of variance (ANOVA), which confirms project aesthetics as indeed a significant predictor of contractor referral ($F_{(1, 114)} = 216.232$, $p = 0.00$).

The analysis also yielded regression equations to support the significance of aesthetics (As) to client satisfaction. With respect to the two dependent constructs of repatronage (Rp) and referral (Rf), the mathematical relationships established are:

$$Rf = 0.9435*As + 0.4193 \tag{1}$$

$$Rp = 0.6101*As + 18.985 \tag{2}$$

These relationships show that any increase or decrease in the aesthetic component of a constructed road would have a direct effect on client satisfaction and by extension contractor referral and repatronage. Contractors are therefore led to prioritise the activities relating to aesthetics in order to ensure that their clients are satisfied with road projects specifically, and other construction projects generally. Although all attributes of quality contribute to client satisfaction, the weightings of their influence differ. Thus, whilst clients demand reliable, highly performing roads, contractors need to deliver to very high aesthetic standards. If

Table 9.2 Regression ANOVA[a]

Model		Sum of Squares	Df	Mean Square	F	Sig.
1	Regression	20987.419	1	20987.419	216.232	.000[b]
	Residual	11064.822	114	97.060		
	Total	32052.241	115			
2	Regression	21592.097	2	10796.048	116.692	.000[c]
	Residual	10460.145	113	92.568		
	Total	32052.241	115			

Notes:
a. Dependent Variable: Referral
b. Predictors: (Constant) Aesthetics
c. Predictors: (Constant) Aesthetics, Performance

contractors can perform well and especially make their construction projects aesthetically appealing, then clients will most likely remain happy. The implications of these research findings are elaborated in the discussion below.

Discussion

Construction processes should lead to the satisfaction of clients and other stakeholders. The three elements of project quality, project performance, reliability and aesthetics, seem to influence clients' satisfaction with constructed roads in Nigeria. Aesthetics was in particular a significant contributor to both the repatronage and referral of contractors, as shown in Figure 9.2. Whereas both performance and reliability are embedded in the construction process, physical appearance of the construction product seems to weigh heavily on the satisfaction derived from it.

Achieving high aesthetics in a project entails the implementation of quality standards, the presence of excitement factors and the durability of the finishing employed, as well as health and safety considerations for users of the constructed road. The facets of design, establishing specifications and carrying out construction and maintenance, all play a role in the final aesthetics. However, most users may not be involved in these aspects of project delivery, and thus base their satisfaction

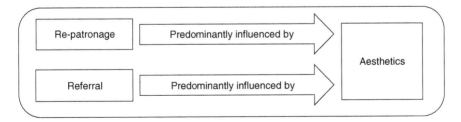

Figure 9.2 The predominant influencer of contractors' repatronage and referral in road projects in Rivers State, Nigeria

on the final appearance and functioning of the product. An adequate maintenance culture is thus worthwhile as it means maintaining the physical appearance and functioning capacity of constructed roads as both clients and road users hinge their continuing satisfaction on these.

From the view of the clients who participated in this research, aesthetics is the most important factor influencing both repatronage and referral. In contrast, contractors and other stakeholders indicated that reliability and performance were their main driving factors of satisfaction (Obunwo, 2016). These results may be linked to the fact that most clients who procure the road projects involved in the study are interested in using physical looks (aesthetics) to demonstrate their provision of good public services and infrastructure. While appreciating the place of aesthetics, clients (who in this research are mainly government representatives) should ensure they also represent the interests of the end users of the roads towards performance and reliability. Clients should thus not ignore the durability and other performance factors which affect the usability of their constructed roads.

The search for satisfaction for the end users of roads should be given importance throughout the project life cycle. This will ensure that completed projects lead to high and acceptable levels of satisfaction. Roads and other construction projects involve activities embedded in engineering (civil, electrical, structural, etc.), land and quantity surveying, architecture and construction project management, amongst others. These activities all have their key performance indices and requirements, which can be guide marks, for the ultimate satisfaction of clients.

Admittedly, project quality alone is not responsible for client satisfaction in road and other types of construction projects. Nevertheless, achieving high or the highest levels of project performance, reliability and especially aesthetics are a step in the right direction for firms and construction stakeholders who intend to optimise construction activities and ensure that the intended clients are satisfied with the constructed roads.

Conclusions and recommendations

This chapter contributes to delivering on the CIB W118 research roadmap by looking at satisfaction from the perspective of Nigerian clients. The findings of the research suggest that construction stakeholders should focus their attention on the attributes of performance, reliability and especially aesthetics in order to deliver durable roads that meet the commuting and transportation needs as well as the satisfaction requirements of clients. Achieving these attributes in full would involve many things, including: prudent cost management, innovative project designs, organisational team work, project supervision, employee training, avoidance of project abandonment, employee motivation, clear design specifications, use of professional experts, proactive maintenance culture, benchmarking, constant inspection of the construction processes, implementation of quality standards, presence of excitement factors and durability of finishing employed as well as health and safety considerations. These activities are geared at enhancing the construction process and ensuring that constructed projects, in this case roads, meet the satisfaction expectations of the concerned stakeholders.

Though the results of our study may not reflect the perspective of developed countries, the results are applicable to the satisfaction of clients in developing countries where their construction industries have contexts similar to Nigeria's. Significant points from this chapter have been summarised below as the implication of client satisfaction regarding construction policies.

For contracting companies to cope with increasing competition, there needs to be a guided focus on client relationships and the satisfaction of the users of their construction products. Adopting various strategies such as enhancing the quality of the constructed project leads to organisational benefits which include improved communication between stakeholders, evaluating construction progress as well as monitoring accomplished results and changes. The identification of project aesthetics as a valuable source of client satisfaction is an indication of the clients' expectation from constructed projects. Whereas these projects require structural rigidity and adherence to design specifications, both clients and users may not necessarily be knowledgeable in the technical details of construction activities. Hence, their assessment of the constructed project is based on what they can see, touch or feel and use in comparison with their earlier expectations and it is the responsibility of contractors to ensure that these feelings are positive. This is in agreement with the earlier definition of satisfaction as a psychological feeling when expectations are either met or exceeded and this can only be measured from the use of the physically constructed project.

Contractors should aim to meet the expectations of their clients and if possible exceed these expectations as this would lead to repatronage and even referrals while sustaining their competitive advantage over competitors. Providers of public facilities especially should thus be wary of just delivering client satisfaction. They should also be proactive at meeting their clients' expectations and be strategic about it. They can effectively use benchmarking to implement and track client satisfaction.

The conclusions and recommendations can be summarised as:

- Clients are particularly interested in the aesthetics of construction projects. However, the performance and reliability of the construction project are also essential. Collectively the trio give a broader understanding of the quality of the constructed project.
- From the regression equations undertaken above, any increase (or decrease) in the aesthetics of construction outcomes would yield a concomitant change in the level of client satisfaction obtainable.
- Project quality is essential to achieving client satisfaction. Its attributes of performance, reliability and aesthetics need to be optimised to achieve greater client satisfaction.
- Contractors are responsible for delivering (public) construction projects. Hence they need to implement measures that would lead to their repatronage or referral based on good past performance.

- The client is the owner of the construction project and is at the centre of affairs pertaining to the construction stakeholders. Clients should thus endeavour to reflect the needs of users in their considerations.

As clients will be more satisfied when the design and construction of their roads are aesthetically pleasing, the result of this research has a number of implications for policy development in Nigeria. Clients of road construction projects will also be satisfied when the finished roads are reliable and perform considerably well. This research recommends that, though aesthetics is a very high predictor of client satisfaction, policy makers should also focus attention on performance and reliability in order to satisfy the end users more. Metrics for monitoring the performance of Nigerian roads over time should be considered. The policy should also factor in the making of Nigerian standards for road design and construction to match internationally recognised standards.

References

Adenugba, A. A. and Ogechi, C. F. (2013). The Effect of Internal Revenue Generation on Infrastructural Development: A study of Lagos State Internal Revenue Service. *Journal of Educational and Social Research*, Vol. 3 (2), 419–36.

Al Nahyan, M. T., Sohal, A., Fildes, B. and Hawas, Y. (2012). Transportation infrastructure development in the UAE: Stakeholder perspectives on management practice, *Construction Innovation: Information, Process, Management*, Vol. 12 (4), 492–514.

Ali, H. A. E. M., Al-Sulaihi, I. A. and Al-Gahtani, K. S. (2013). Indicators for measuring performance of building construction companies in Kingdom of Saudi Arabia. *Journal of King Saud University – Engineering Sciences*, Vol. 25 (2), 125–34.

Aniekwu, N. (1995). The business environment of the construction industry in Nigeria. *Construction Management and Economics*, Vol. 13, 445–55.

Atkinson, A. A., Waterhouse, J. H. and Wells, R. B. (1997). A stakeholder approach to strategic performance measurement. *MIT Sloan Management Review*, Vol. 38 (3), 25–37.

Beatham, S., Anumba, C., Thorpe, T. and Hedges, I. (2004). KPIs: a critical appraisal of their use in construction. *Benchmarking: An International Journal*, Vol. 11 (1), 93–117.

Boyd, D. and Chinyio, E. (2006). *Understanding the construction client*. Oxford, UK: Blackwell Publishing Ltd.

Cheng, J., Proverbs, D. G., Oduoza, C. and Fleming, C. (2005). A conceptual model towards the measurement of construction client satisfaction. In: Khosrowshahi, F. (Ed.). *Proceedings of 21st Annual Association of Researchers in Construction Management (ARCOM) Conference, Held at the SOAS University of London (School of Oriental and African Studies), 7–9 September*. Vol. 2, 1053–62.

Eadie, R. and Graham, M. (2014). Analysing the advantages of early contractor involvement. *International Journal of Procurement Management*, Vol. 7 (6), 661–76.

Gwilliam, K., Foster, V., Archondo-Callao, R., Briceño-Garmendia, C., Nogales, A. and Sethi, K. (2009). *The burden of maintenance: roads in sub-saharan Africa*. AICD Background Paper 14. Africa Region, World Bank, Washington, DC.

Haugbølle, K. and Boyd, D. (2013). *Clients and Users in Construction: Research Roadmap Report*. CIB Publication 371. Rotterdam: CIB General Secretariat. Available at: http://site.cibworld.nl/dl/publications/pub_371.pdf (Accessed 2 May 2016).

Idoro, G. (2010). Influence of quality performance on clients' patronage of indigenous and expatriate construction contractors in Nigeria. *Journal of Civil Engineering and Management*, Vol. 16 (1), 65–73.

Jonsson, A. and Svingby, G. (2007). The use of scoring rubrics: Reliability, validity and educational consequences. *Educational Research Review*, Vol. 2 (2), 130–44.

Karna, S. (2014). Analysing customer satisfaction and quality in construction: The case of public and private customers. *Nordic Journal of Surveying and Real Estate Research – Special Series*, Vol. 2, 67–80.

Koster, K. (2010). *International project management*. London: Sage Publishers.

Love, P. E. (2002). Influence of project type and procurement method on rework costs in building construction projects. *Journal of Construction Engineering and Management*, Vol. 128 (1), 18–29.

Masrom, M., Skitmore, M. and Bridge, A. (2013). Determinants of contractor satisfaction. *Construction Management and Economics*, Vol. 31 (7), 761–79.

NPC (2012). *The Nigerian Economy: Annual Performance Report*. Abuja: The Presidency, National Planning Commission Report.

Obinna, V. C., Owei, O. B. and Mark, E. O. (2010). Informal settlements of Port Harcourt and potentials for planned city expansion. *Environmental Research Journal*, Vol. 4 (3), 222–8.

Obunwo, C. (2016). *A framework for enhancing project quality and customer satisfaction in Government road construction projects in Rivers State, Nigeria. PhD Thesis.* Wolverhampton: University of Wolverhampton.

Obunwo, C., Chinyio, E. and Suresh, S. (2013). Quality management as a key requirement for stakeholders' satisfaction in Nigerian construction projects. In: Ahmed, V., Egbu, C. O., Underwood, J., Lee, A. and Chynoweth, P. (Eds.). *Proceedings of the 11th International post graduate research conference, 8–10 April 2013, Held at the school of the Built Environment, University of Salford, Manchester*, pp. 723–34.

Odediran, S. J., Opatunji, O. Y. and Eghnure, F. O. (2012). Maintenance of residential buildings: Users' practices in Nigeria. *Journal of Emerging Trends in Economics and Management Sciences*, Vol. 3 (3), 261–5.

Ogunlana, S. O. (2010). Beyond the 'iron triangle': Stakeholder perception of key performance indicators (KPIs) for large-scale public sector development projects. *International Journal of Project Management*, Vol. 28 (3), 228–36.

Olatunji, A. and Diugwu, I. A. (2013). A project management perspective to the management of federal roads in Nigeria: A case study of Minna–Bida road. *Journal of Finance and Economics*, Vol. 1 (4), 54–61.

Oluwakiyesi, T. (2011). *Construction Industry Report: A Haven of Opportunities*. Vitiva Research, available at www.proshareng.com/admin/upload/reports/VetivResearch ConstructioSectorReportMay2011.pdf (Accessed 3 January 2013).

OPEC (2010). *World Oil Outlook* [online]. Available at www.opec.org/opec_web/static_files_project/media/downloads/publications/WOO_2010.pdf (Accessed 21 May 2016).

Pheng, L. S. and Chuan, Q. T. (2006). Environmental factors and work performance of project managers in the construction industry. *International Journal of Project Management*, Vol. 24 (1), 24–37.

Pritchard, M. P., Funk, D. C. and Alexandris, K. (2009). Barriers to repeat patronage: The impact of spectator constraints. *European Journal of Marketing*, Vol. 43 (1/2), 169–87.

Rahman, A. and Alzubi, Y. (2015). Exploring key contractor factors influencing client satisfaction level in dealing with construction projects: An empirical study in Jordan.

International Journal of Academic Research in Business and Social Sciences, Vol. 5 (12), 109–26.

Rhodes, C. (2015). Business statistics. Economic policy and statistics. House of Commons Briefing papers SN06186.

Sanni, A. A. and Windapo, A. O. (2008). Evaluation of contractors' quality control practices on construction sites in Nigeria. In: Dainty, A. (Ed. 2008). *Proceedings of the 24th Annual ARCOM Conference, 1–3 September 2008, Held in Cardiff, UK, Association of Researchers in Construction Management*, pp. 257–65.

Soetanto, R. and Proverbs, D. G. (2001). Modelling client satisfaction levels: A comparison of multiple regression and artificial neural network techniques. In: Akintoye, A. (Ed. 2001). *Proceedings of 17th Annual ARCOM Conference, 5–7 September 2001, University of Salford. Association of Researchers in Construction Management*, Vol. 1, 47–57.

Torbica, Z. M. and Stroh, R. C. (2000). HOMBSAT: An instrument for measuring home-buyer satisfaction. *Quality Management Journal*, Vol. 7 (4), 32–44.

UKCG (2012) Importance of construction to industry. Available at www.ukcg.org.uk/representing-industry/creating-britains-future/importance-of-construction-to-industry/ (Accessed 22 February 2017).

UN (2014). *World Urbanization Prospects: UN Department of Economic and Social Affairs*. New York: United Nations. Available at http://esa.un.org/unpd/wup/Publications/Files/WUP2014-Highlights.pdf (Accessed 2 May 2016).

UN-Habitat (2014). *PrepCom1 Summary Compilation*. New York: UN Habitat. Available at www.esa.un.org/. (Accessed 2 May 2016).

Vennstrom, A. and Eriksson, P. E. (2010). Client perceived barriers to change of the construction process. *Construction Innovation*, Vol. 10 (2), 126–37.

Weil, D. (2005). The contemporary industrial relations system in construction: Analysis, observations and speculations. *Labor history*, Vol. 46 (4), 447–71.

Whittaker, G., Ledden, L. and Kalafatis, S. P. (2007). A re-examination of the relationship between value, satisfaction and intention in business services. *Journal of Services Marketing*, Vol. 21 (5), 345–57.

Xiong, B., Skitmore, M., Xia, B., Masrom, M., Ye, K. and Bridge, A. (2014). Examining the influence of participant performance factors on contractor satisfaction: A structural equation model. *International Journal of Project Management*, Vol. 32 (3), 482–91.

Part III
Innovation
Change versus stability

10 Stimulating innovation through integrated procurement

The case of three-envelope tendering

Ada Fung and Ka-man Yeung

Introduction

As pointed out by the CIB W118 research roadmap (Haugbølle and Boyd, 2013) there is a pending need to consider the ways clients procure services and products from the construction industry. Procuring services and products in a way that may also stimulate innovation in and by the construction industry is by no means a small task. Rather, developing new ways of procuring services and products in order to stimulate innovation is a challenge with large potential ramifications. In recent years, the Hong Kong Housing Authority has developed and tested an innovative new contracting system called the three-envelope tendering system.

The Hong Kong Housing Authority (HA) is a statutory body in Hong Kong and is financially autonomous, with the Housing Department (HD) working as its executive arm. The HA is responsible for producing and managing the public housing programme in Hong Kong and is one of the largest in the world, helping low-income families gain access to affordable rental housing and lower-middle-income families to own their homes through subsidised home ownership schemes. New and comfortable living environments are created as a hub for the community by HA housing developments. Housing estates embrace social and retail facilities and amenities besides residential properties. The HA has currently over 160 public rental housing estates serving about 744,000 families under its care across Hong Kong, built in different eras. It is essential that the HA has to build for the future: quickly, efficiently and in ways that optimise the space available and are fully sustainable (Hong Kong Housing Authority, 2015). To this end, the HA has been conducting research and development (R&D) activities continuously in order to improve its delivery of public housing, aiming to achieve a more cost-effective use of resources, drive innovations and unleash people potential in face of the huge demand for public housing with rising aspiration for quality living.

This chapter aims to illustrate how innovative designs and construction approaches within affordability can be obtained from the industry stakeholders through innovative contract procurement, using so-called three-envelope tendering. This chapter will describe the HA's procurement system and the considerations guiding the development of its three-envelope tendering system for integrated contracts, and case studies will illustrate the benefits obtained from the implementation of the system in projects.

The HA's procurement system

The procurement system for construction projects is very much concerned with the organised method by which the project owner will obtain, acquire or bring about its desired construction products. It is a process of identification, selection and commissioning of the contributions required for the development of the project. This reflects the organisational arrangement (e.g. the process, procedures, activities sequencing and organisational approach in project delivery) and the contractual arrangements (e.g. the allocation of risks and responsibilities), which will determine the project success and can be made to ensure that the appropriate contributions are properly commissioned and the interests of the project owner are safeguarded (Abdul Rashid et al., 2006).

Being a statutory body established for producing and managing the public housing programme in Hong Kong, the HA procures high volumes of building construction works every year. For the construction of public housing developments, the HA generally adopts the conventional design–bid–build procurement model. The HD, as the executive arm of HA, is equipped with strong in-house professional teams, and maintains full responsibilities for design management and contract management throughout the delivery process. This includes the preparation of the design and tender documents for construction contracts to procure contractors for execution of construction work. Occasionally, external professional services providers are engaged to supplement or complement in-house resources to assist in the provision of design services.

Since the HA does not possess a construction services arm, all construction works are executed by qualified contractors procured through selective competitive tendering conforming to the Government Procurement Agreement of the World Trade Organisation (WTO GPA). The contractors are fully responsible for carrying out the works in accordance with the design and specification prescribed by the design team in the contract documents, and are required to maintain continuous supervision on site at all times until completion of the works and handing over to the HA. Contract administration duties are taken up by contract managers who are chief professionals of HD, and they are responsible for periodic supervision of the contractors' work on site. In respect of specialist building services works, nominated sub-contractors are nominated by the HA to the contractor to carry out these works. In each foundation and building contract, the Dispute Avoidance and Resolution Adviser (DARA) system is adopted to help foster cooperation among the contracting parties and project team members and to facilitate dispute avoidance and prompt resolution of disputes. The essence of this organisational approach is that the HD appoints a multidisciplinary design and project management team with the architect/engineer as the contract manager also be acting as the principal adviser and agent for the HA, and to lead

Public housing development in Hong Kong

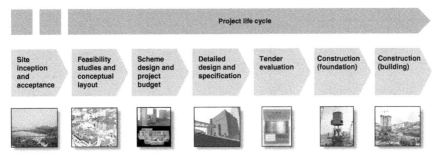

Figure 10.1 The project life cycle of public housing development in Hong Kong

the process throughout the project life cycle (Figure 10.1). The projects proceed in a linear manner, with construction following design and tendering.

The merits of the conventional design–bid–build procurement arrangement are that the contract manager and his or her design team and the contractor are all conversant with the process and their well-established roles, and the HA has total control over the design and quality of the project. Commitment to spend large sums of money under the construction contracts comes at a relatively late stage of the project and the tendering on substantially completed design information offers a reasonably high level of certainty of cost, quality and programme before committing to build. Such procurement arrangement has been operating effectively with low risks and provides high reliability to the HA (Masterman, 2002).

In respect of procurement of construction projects, the HA generally adopts selective tendering by using the various standing lists of qualified contractors which it has established and maintained. The contractors are prequalified and admitted to the lists based on their financial, technical and management capability and relevant work experience. The HA maintains at all times on each standing list a sufficient number of qualified contractors to ensure that competitive tendering may be conducted for each tender. To provide an incentive for good performers, shortlisting of contractors is conducted quarterly to determine their tender opportunities in the following quarter based on their performance measured under the HA's Performance Assessment Scoring System (PASS).

For small, simple and straightforward construction contracts, tenders are submitted and assessed on a single envelope basis. The submitted tender together with the completed tender price documents are evaluated based on price and past performance with a price to non-price ratio of 75:25 for tender assessment.

For relatively complex capital works building and foundation contracts, a two-envelope tendering system is adopted in which each tender is submitted in two envelopes, that is, an envelope on the technical proposal and another envelope on the price proposal for execution of the works. The submitted tenders are evaluated on price and quality assessments, an evaluation process combining the assessment of the tender price and the technical merits of the tenderers by taking

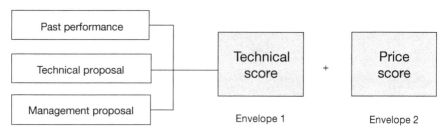

Figure 10.2 Tender evaluation in the two-envelope tendering system

into consideration their past performance, and the technical and management proposals of the tenderers (Figure 10.2). The respective weights for price and quality would be determined on account of the level of quality of services required, risk consideration, technical expertise, time constraint, complexity or special features of the contract. The price to technical scoring ratio generally ranges from 80:20 to 65:35. The criteria for awarding the contract, including the details of the marking scheme and all factors to be considered in the evaluation of tender, are stipulated in the tender documents and made known to tenderers. The contract will normally be awarded to the tenderer of the conforming tender assessed with the highest combined technical score and price score.

The three-envelope tendering system for integrated contracting

Catalyst for a new system: Kai Tak Site 1B

In 2008, the HA was tasked to develop a large site of 5.7 hectares at Kai Tak Site 1B at the former airport stripe in Hong Kong harbour. The new housing area should include 8,100 new public rental housing flats and a greening ratio of over 30 per cent to provide housing accommodation for about 20,000 people with the first tenants moving in during 2013. The site had previously been installed with concrete piles for a home ownership scheme which was abandoned in 2002.

The main challenges and opportunities with this development included:

- a tight development programme for the large production quantum;
- the need to maximise the use of the installed piles to reduce abortive piles and minimise their adverse implications for the construction of the new housing blocks;
- the fulfillment of newly introduced sustainability and environmental requirements;
- aspirations to enhance and advance prevailing design and construction practices;
- tight in-house professional and technical resources to support the design and management of the project through to the obtainment of the occupation permit; and

- a tight budget to afford the use of the traditional two-envelope tendering system with high financial risk associated with innovations.

In general, the traditional design–bid–build procurement arrangement and the prevailing tendering system have been operating satisfactorily for HA projects. But the HA recognised that this arrangement, together with the prevailing tendering system adopted in its contracts, were not prudent enough to meet the challenges and opportunities of this project, and there was imminent need for an innovative procurement system as a solution to help improve the environmental friendliness and sustainability of the design, site productivity and built quality for the delivery of this monumental public housing project.

Formulation of the three-envelope tendering system

To meet the challenges and the complex demands of the development, a working group was formed in the HD to explore a suitable procurement system solution. This was led by the deputy director of Housing (Development and Construction) and comprised senior management and professionals of the development and construction division of the HD. The working group reviewed the strengths and weaknesses of the existing HA procurement system in conjunction with the objectives of the development, brainstormed possible issues and options, and consulted industry stakeholders and experts widely for their views. The working group concluded that there was no readily available procurement system that could fit both the requirements of the HA and the project. From the feedback received from project teams, the working group agreed that the prevailing procurement system of the HA was well practised with proven reliability and a good track record so it should not be simply abandoned to a brand new system that was unfamiliar and not tested. Besides, consideration should be given to public aspiration on the HA in the project delivery. On balance, the working group held the view that the HA as a public body should maintain a high level of participation throughout the project delivery without being perceived as shifting the work and responsibilities to its business partners.

To enable its further study and research for appropriate options, the working group established and adopted the following guiding principles for its work:

- The fair and equitable risk-sharing principles practiced by the HA in its procurement systems should be upheld as they were well received by the industry.
- The prevailing modus operandi for design management and contract management was effective and should be maintained as far as possible.
- Opportunity for earlier and better integration of design and construction expertise from employer, designers, builders and manufacturers should be explored so as to achieve synergistic results for the provision of a holistic and environmentally friendly design solution, practising the 'lean construction'

concept (Koskela *et al.*, 2002), eliminating waste and abortive work in the delivery process.

- Encouragement for innovations from business partners and transference of the innovations for general application in future HA's projects within an acceptable risk margin to the HA should be incorporated where practicable.

After a series of preparations, discussions and consultations, the working group created an innovative three-envelope tendering system which would operate alongside the traditional procurement system, and this was adopted by the HA in 2009. An integrated contract for the design and construction of this large-scale public housing development was procured using the three-envelope system. The integrated contract with integrated procurement approach was developed as a hybrid procurement mode designed to suit the HA's risk appetite, featuring benefits from both the HA's 'design–bid–build' and the industry's 'design–build' procurement modes. It facilitated the development of an integrated design and construction process and the application of the concept of sustainable development through research and innovation. This contract integrated the client's project team and the contractor and its design and construction counterparts to work collaboratively for timely achievement of the production targets. Contractors' expertise in construction works were engaged upfront at the tendering stage whereby the tenderers were required through tendering conditions to submit innovation proposals as well as the design/technical and price proposals.

This integrated procurement approach facilitated the effective use of external resources and enabled earlier and better integration of the contractor's expertise into the design of the works. The HD's professional team played an active role throughout the process, from procurement to completion. Based on their familiarity with the client's requirements and experience in the HA's design practices, the professional team prepared a schematic base plan based on the client's requirements and provided it to the tenderers as reference and for illustration of the client's intentions and requirements. This showed the overall planning and zoning of various parts of the site, the design of the domestic blocks, retail and car-parking facilities, choice of materials and workmanship, and any specific requirements for green and innovative design features. This helped ensure that the tenderers could properly understand the contract requirements.

The tender documentation prepared by the HD's professional team included the following:

- The client's requirements, which stipulated in detail the client's brief, programme and phased completion of the works, design standards and specifications, detailed design and other requirements.
- A base scheme which provided an indicative design that could meet the fundamental aspects of the client's requirements for tenderers' reference. It was not meant to be an exhaustive and finalised design. The tenderers maintained full discretion for their design to be offered.

- The technical submission requirements, tender evaluation criteria and scoring scheme.

The integrated contract embraced all specialist works, such as piling works, building services installation works and soft landscape works, and all design services, comprising architectural, building services, structural and geotechnical engineering services, in one single contract. For the sake of quality assurance, and to minimise the risks exposure for all parties, the tender documents provided tenderers with lists of prequalified and shortlisted specialist sub-contractors and design consultants specifically compiled by the HA for this project. The tenderers were permitted to team up with the specialist sub-contractors and consultants as domestic sub-contractors and design consultants selected from such lists for tendering the integrated contract.

For competitive tendering, the tenderers were required to make their offers based on a common set of tender parameters. Deviations from these tender parameters could be regarded as disqualifying tenders and render the submitted tenders invalid. To make room for tenderers' submission of proposals for tapping their innovative ideas and construction expertise, it necessitated the development of new tender parameters that could facilitate submission of tenderers' proposals which would not have been detailed nor specified in the tender documents whilst the client had the flexibility to consider and adopt any such submitted proposals that fitted the time, cost and quality objectives of the project. On this basis, the three-envelope tendering system was created for integrated contracting. Under this system, tenderers were required to submit their tenders comprising:

- technical proposals on proposed designs and method statements;
- innovation proposals including associated price adjustment due to each innovation proposal item therein; and
- price proposal for execution of the works.

The concept of the three-envelope tendering system is illustrated in Figure 10.3.

The three envelopes

Under the innovative three-envelope tendering system, the tenderers submitted their tenders in three envelopes as follows:

- Envelope 1: corporate information and technical submission conforming to the client's requirements.
- Envelope 2: proposals for innovation which were over and above the client's requirements, and associated price adjustment for each proposal.
- Envelope 3: price for all works in full compliance with the client's requirements.

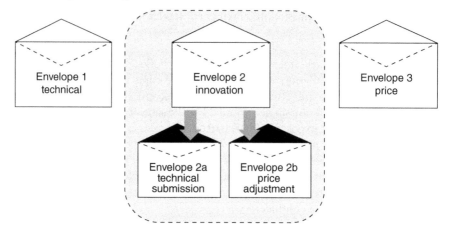

Figure 10.3 The three-envelope tendering system

A comparison of the conventional two-envelope tender submission arrangement with the three-envelope submission system is shown in Figure 10.4.

Envelope 1 comprised two smaller envelopes: sub-envelope 1a and sub-envelope 1b. Corporate information of the tenderers and the proposed management and technical resources were enclosed in sub-envelope 1a wherein the tenderers' name and key persons were revealed in the submissions for technical assessment. The sub-envelope 1b needed to be enclosed with the design and technical information with proposal for master planning, proposal for detailed design, proposal for programme of works and construction management, and proposal for sustainable construction. To ensure impartial and fair assessment at tender evaluation stage, all tenderers were restricted from showing or indicating any description, logo or graphic that would reveal their identity in their design and technical information included in the sub-envelope 1b.

For envelope 2, the tenderers could make research and offer innovation proposals without limitation on the number or the total amount of proposals to be offered at their discretion. The innovation proposals to be submitted needed to comply with the following five criteria:

- They were not specified requirements.
- They had not been implemented in HA projects or only implemented in some HA projects but not in full scale.
- They were technically justified.
- They had the potential for wide application in other HA projects.
- They had local project references. If not, at least experience outside Hong Kong should be provided for consideration.

Each tenderer was required to provide a list of innovation proposals and submit the innovation proposals in envelope 2, comprising the technical submissions in sub-envelope 2a and the corresponding price adjustments for individual proposal

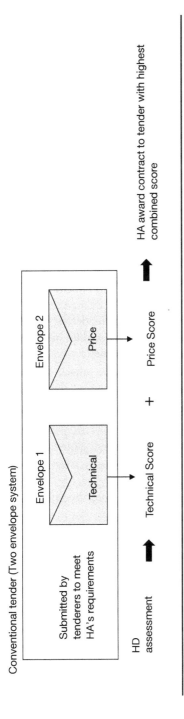
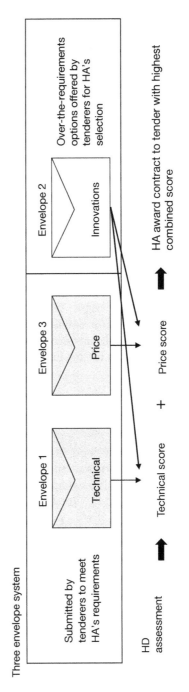

Figure 10.4 Two-envelope tendering system versus three-envelope tendering system

of innovation in sub-envelope 2b. The information in envelope 2 was also submitted anonymously.

In envelope 3, the tenderers were required to submit the price information for all works (but excluding innovation proposals) in full compliance with the client's requirements. The price information should include a breakdown of the contractor's rates and prices in sufficient details with approximate quantities showing the build-up of the total sum for the works. Guidelines for preparation of the breakdown of the contractor's rates and prices and a model schedule of rates were included in the tender documents which stipulated, among others, the minimum requirements on the level of details and format of the price information to be submitted.

In order to maintain the integrity in the assessment process, the technical assessment of the tenderer's technical proposals was conducted in anonymity, and the information in sub-envelopes 1b, 2a and 2b was not allowed to reveal the tenderers' identity. The tender conditions required each tenderer to create his or her own tenderer's identity number (TIN) by using a 6-digit figure which was to be clearly stated on each and every page of the anonymous documents. Each tenderer was also required to provide a sealed envelope with a statement clearly revealing the tenderer's identity of the TIN and the sealed envelope was to be submitted under the sub-envelope 1a. The HD formed a dedicated team separate from tender evaluation teams to conduct screening of the tenderers' submissions to ensure that tenderers' identities were not shown or were duly hidden before they were passed to the tender evaluation teams for assessment.

The tender evaluation process

The tender evaluation was separated into two stages:

- In Stage 1, envelopes 1 and 2 were opened and the tenderers' submissions were assessed and scored, including the selection of proposals for innovation, to come up with the non-price technical score.
- In Stage 2, after HA's endorsement of the non-price score, the price envelope (i.e. envelope 3) was opened and the total price based on the tender price offer in Envelope 3 together with the price adjustments due to the selected proposals of innovation from envelope 2 were calculated.
- Then, the combined scores of the respective tenders were worked out based on the total price and non-price scores of the tenders and this was used as the basis for consideration of award of the contract (Figure 10.5). The design and technical submissions of the successful tenderer, including those selected proposals for innovation, would form part of the contract.

For better accountability, tender assessment was conducted by separate professional teams from the HD with a two-tier mechanism as follows:

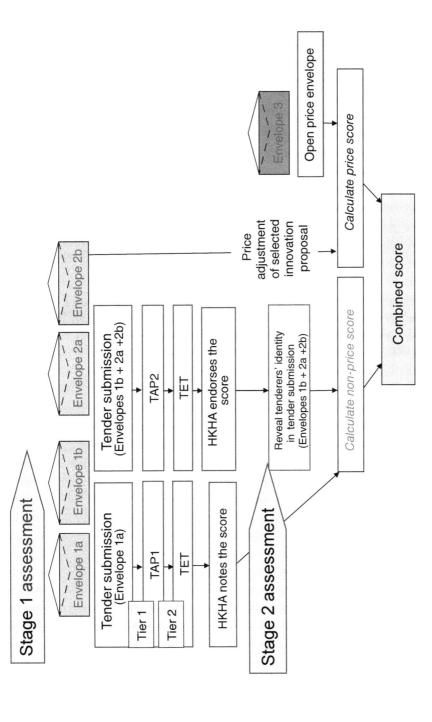

Figure 10.5 Two-stage and two-tier tender assessment process in the three-envelope tendering system

- Tier 1 assessment: two independent technical assessment panels (TAP) comprising professional teams of the HD were formed. The first TAP examined the technical submission of corporate information of the tenderers and the proposed management and technical resources in the sub-envelope 1a with the names of the tenderers revealed. The second TAP examined the design and technical submission in sub-envelope 1b and envelope 2 which remained anonymous, with proposals for master planning, detailed design, programme of works and construction management, any proposals for sustainable construction and innovation proposals with stipulated evaluation criteria. The TAPs would then make a recommendation to the tender evaluation team on the technical assessment.
- Tier 2 assessment: a tender evaluation team (TET) comprising directorates of the HD were formed to evaluate the technical assessment results recommended by the two independent TAPs, score the technical submissions and select the innovation proposals according to the stipulated evaluation criteria and scoring scheme. Upon endorsement of the technical scores by HA, the TET should open the price envelope, i.e. envelope 3, and work out the combined scores.

The workflow of the two-stage two-tier tender assessment process for the three-envelope tendering system is shown in Figure 10.5.

To strike a balance between prudent cost control and quality design with innovation to obtain the best value for money, the weighting for the non-price to price–score ratio was set at 45:55 in which the weighting of the non-price elements was heavier than that in the prevalent two-envelope tendering system used for other contracts of the HA (see Figure 10.6).

Assessment and selection of innovation proposals

An innovation proposal was defined as a proposal which was not a specified requirement in the contract specification and which had potential for wider application in other HA projects. Each innovation proposal submitted by the tenderers would be assessed in respect of their benefits to be brought about by the proposal and classified in three different benefit levels, namely high, medium and low, and a corresponding technical score would be allotted according to its benefit level. To assist the tenderers to work on areas of merits by providing them with reference and guidance, the tender documents included:

- a list of initiatives that have been tried out or adopted in previous HA projects which would not be considered as innovations if no major improvement was proposed by the tenderer. Examples included the use of self-compacting concrete, precast structural walls and precast stair core;
- a list of initiatives that could be considered as innovations of low benefit level if proposed by the tenderer. Examples included the use of precast slabs, precast facades of 2-storey height, reuse of lift re-generative power; and

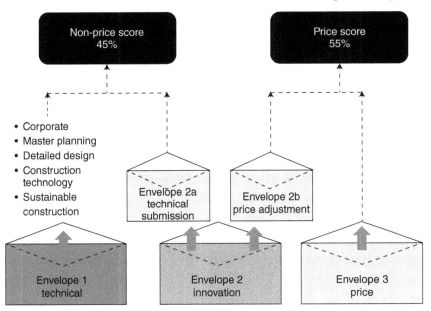

Figure 10.6 The price to non-price ratio of three-envelope tendering system

- a list of initiatives that might be considered as innovations. Examples included any methodology for recycling at least 30 per cent of construction waste, utilisation of 5 per cent or more of reusable materials and reduction in annual sewage volumes by 20 per cent or more.

To maintain a reasonable balance within the tenders for innovation proposals, the total allotted score for all selected proposals would not exceed 5 marks out of the total non-price score at 45 marks and the sum of corresponding price adjustments would not exceed a pre-set ceiling amount of HKD 200 million (EUR 23.4 million).

Each tenderer was required to provide a list of innovation proposals with his or her tender. For each proposal of innovation or initiative, information had to be submitted showing:

- the detailed proposal together with information for justification, such as job reference, operational and maintenance manual, method statement, specifications on material/equipment, workmanship and testing requirements for any new materials/equipment proposed, details of verification test, mock-up or trial, test reports, evaluation on performance, approval given by relevant authority;
- The service life of the proposal under normal operation conditions;
- service lives of major components of the proposal under normal conditions;
- the cost of supply and construction of the proposal;
- the cost of routine maintenance of the proposal within its service life;

- the cost of replacement of major components within the service life of the proposal;
- the cost of operation of the proposal (including cost of electricity, water, gas, consumables, labour, attendance) within its service life;
- the cost of disposal of the proposal at the end of its service life;
- a warranty for the proposed innovation or initiative; and
- the net cost adjustment to the tender sum and the breakdown of omissions and additions in quantities and in rates/prices for the proposal.

The innovation proposals submitted by the tenderers were assessed, scored and selected through a three-step mechanism which involved:

- opening sub-envelope 2a and assessment of the technical submission of the innovation proposals. Any proposal items that did not comply with one or more of the five criteria for innovations as stipulated in the tender documents were rejected for further consideration. Remaining proposal items were assessed in respect of the benefits to be brought about by the proposal and classified under three different benefit levels, namely high, medium and low and a corresponding technical score according to its benefit level would be allotted to the tender.
- opening sub-envelope 2b and assessment of the submitted price adjustments for the innovation proposals. Any proposed items with the corresponding price adjustment being considered unreasonably high were rejected for further consideration. Remaining proposal items were prioritised in the order of their price adjustments within each benefit level group (i.e. the proposal with the lowest price adjustment was accorded the highest priority) unless there was a distinctively better cost–benefit performance for a certain proposal for which its priority would be adjusted accordingly.
- selection of the proposals in order of the priority with particular consideration for their good value for money and benefits to other new projects of the HA (Figure 10.7).

Risk mitigation with the three-envelope tendering system for integrated procurement

As a public body, risk management by the HA was still an important element of concern. The working group understood that the risks associated with the proposals must be fully considered and it was pivotal to successful procurement in having risk management incorporated holistically at the outset. Riding upon the prevailing practice of the HA, the working group adopted the following approaches for risk management:

- Manage risks as an integral part of the procurement process through identification and assessment of risks, development of risk responses, and conducting subsequent monitoring and review.

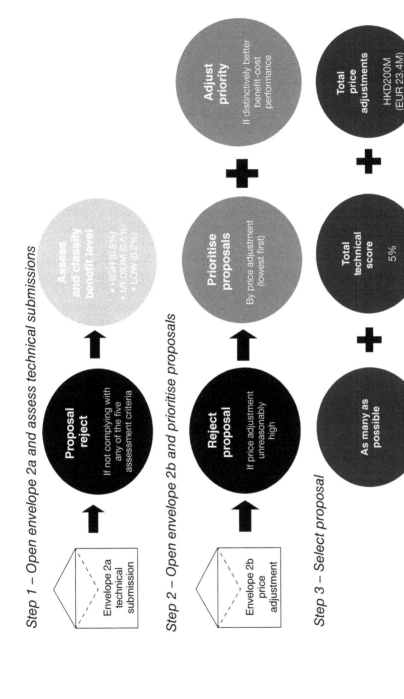

Figure 10.7 Selection of innovation proposals

- Allocate risks to the party who can best manage them.
- Consider risk responses holistically taking into account the overall balance of procurement objectives and principles.

In developing the operational mechanism of the three-envelope system, the working group examined the major risk areas covering aspects of the overall quality of work, design, contractual, financial and operational arrangements, and quantum risks. Risk mitigation measures were suitably put in place as deliberations went on.

For the traditional approach, the tenderers were required to submit tenders based on the design provided by the HA. However, under the integrated procurement with three-envelope tendering system, the tenderers were required to submit tenders based on their designs. This involved high risk to both the HA and the tenderers that the tenderer's resources might not be effectively spent on the production of a competent design meeting the HA's requirements. The working group considered that mitigation measures should be incorporated into the system to minimise the risks on design and overall quality standards:

- HD to prepare a base scheme as an indicative design that met the fundamental aspects of the client's requirements which were issued for tenderers' reference at the tender stage.
- HD also shortlisted professional design consultants and specialist works contractors from the HA's lists or other recognised lists of quality consultants and specialist contractors and included the shortlist in the tender documents for selection by the tendering contractors of the integrated contract.

In order to contain the contractual and operational risks in procurement of the integrated contract, HA engaged a dispute avoidance adviser (DAA) to advise on risk and dispute avoidance during the development and tender documentation preparation processes of this new procurement model. Under the integrated contract, the contractor would also provide professional indemnity insurance and design warranties to the HA for work quality assurance.

The integrated contract with the three-envelope system was a new procurement mode which demanded heavy input from the tenderers during tender stage and entailed high tender cost. The tenderers bore a greater contractual responsibility than the conventional procurement mode and might experience greater financial risk. In order to save resources and ensure effective competition, the working group considered that the number of tenderers for the integrated contract should not be excessive and, taking consideration of the size and complexity of this project, the number of tenderers was kept to three only. To alleviate the financial burden on the tenderers in their preparation of tenders, the HA offered a one-off payment of HKD 1 million (EUR 0.12 million) as honorarium to each of the conforming but unsuccessful tenderers and in return the tenderers allowed the HA the rights to use the design and innovation proposals submitted with their tenders in HA projects.

Implementation of the three-envelope tendering system

This section describes the implementation of the three-envelope tendering system in two of the HA's mega-projects and the experience gained. The characteristics of the two projects are summarised in Table 10.1.

Project 1 – Kai Tak Site 1B

Amidst the outbreak of the global financial tsunami and the fallout of the financial industry in 2008, the Hong Kong gross domestic product (GDP) growth rate fell from +7.3 per cent in the first quarter of 2008 to –2.5 per cent in the fourth quarter (Hong Kong Government, 2009). The Hong Kong economy was

Table 10.1 Factsheet of the two projects implemented with the three-envelope tendering system

Project	Kai Tak Site 1B	Anderson Road Site A and Site B Phases 1 and 2
Location	Kai Tak, Kowloon	Kwun Tong, Kowloon
Development site area	5.7 ha	5.32 ha
Domestic plot ratio	5.51	7.5
Domestic gross floor area	314,070 m²	351,750 m²
Non-domestic plot ratio	0.1	1.5
Non-domestic gross floor area	5,700 m²	70,350 m²
Number of domestic blocks	9	9
Number of flats	8,100	7,198
Occupants (expected)	20,418	21,256
Parking		
Private car parking for domestic	266	184
Private car parking for commercial	12	32
Light goods vehicles parking	35	24
Motorcycle parking	27	66
Recreation		
Local open space	20,418 m²	21,252 m²
Children's play area	1,633 m²	1,700 m²
Table tennis court	2	2
Basketball court	2	2
Badminton court	2	2
5-a-side soccer pitch	1	0
Retail provisions		
Retail provision	2,760 m²	3,100 m²
Wet market	0	73 stalls
Education		
Nursery class and kindergartens	One 8-classroom kindergarten	Two 6-classroom kindergartens
Estate management office	1	1

hard hit in the financial crisis and the construction industry was no exception. The HA launched the development of the mega-project at Kai Tak Site 1B which was planned to provide 8,100 new public rental housing flats for some 20,000 people. The Kai Tak Site 1B contract has been completed progressively from 2013 to early 2014 (Figure 10.8).

At this economic trough, stakeholders in the construction industry welcomed the move and were extremely keen to share their efforts in what was a lifeblood project for them. This provided the HA with the opportunities to review and reform the existing procurement modes and to seek all its stakeholders in the value chain to build a strong business team to respond to the challenges and, if

SITE AREA: 5.70 ha
FLAT NO.: about 8,000 flats
COST: HK$ 2,946M (€344.6M)
FACILITIES: Commercial,
Parking,
Kindergarten

MERIT
• Site is large and self-contained
• Design opportunities
• Ample working space
• Economy of scale
• Planning, political and technical
risks are low

Figure 10.8 The Kai Tak Site 1B project

possible, to drive the industry forward for excellence in value and quality. After a series of consultations and workshops in early 2009, this project and the three-envelope tendering system received overwhelming support and contribution from the three tenderers including their design and construction counterparts. They sourced innovations worldwide and adapted them for use in the Hong Kong environment. There were 88 innovation items proposed by the three tenderers, and after assessment 7 proposals were selected and qualified for acceptance. Taking into account the cost effectiveness, 4 of these proposals were finally accepted for application in the successful tender. They included:

- an application of i-Crete™ system which was an advanced concrete mix design system that optimised concrete strength and workability with reduced cement content;
- an eco rainwater root zone irrigation system for planters on the roof, which could reduce water loss by evaporation in conventional surface irrigation;
- an external wall painting system containing TiO_2 to help reduce air pollution through decomposition of nitrogen oxides by photocatalytic action of TiO_2; and
- precast construction for all ground floor water tanks of domestic blocks.

In the successful tender, the contractor had also provided valuable design input and proposed a good master plan and detailed design of 9 domestic blocks which effectively opened up the estate entrances, mitigated the traffic noise and enhanced air ventilation and natural light when compared with the schematic base plan. This master plan with good disposition of the domestic blocks also provided a cost-effective design with over 70 per cent reuse of the existing piles on the site.

Project 2 – Anderson Road Sites 1 and 2 Phases A and B

A second project involving the production of 7,100 domestic flats used the new procurement and was tendered in 2012 and contracted in January 2013. The Anderson Road project was under construction in 2016 for completion in early 2017. It was deemed suitable for the three-envelope tendering system due to its favourable site size, low planning risk and suitable scope for design and innovation. During consultation in October 2011, the proposal was well received by the industry.

In this project, 106 innovation items were proposed by the three tenderers and after assessment 5 proposals were selected for application in the successful tender which included:

- 5D Building Information Modelling (BIM) for construction management;
- application of BIM-5D to project management;
- application of CU-Structural Soil to tree pits in pavements;
- permeable external pavers using recycled aggregates; and
- 100 years design life for building structures.

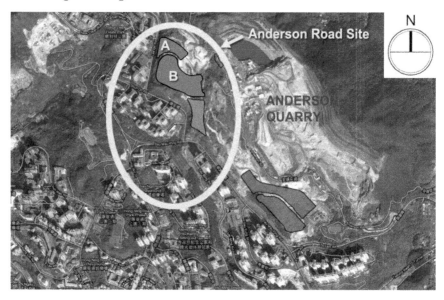

Figure 10.9 The Anderson Road Sites 1 and 2 Phases A and B

Benchmark comparison

As part of the post-completion review, the working group made two benchmarking comparisons of the three-envelope tendering system with various procurement systems used or piloted by the HA:

- comparison of the risk profile of the various procurement arrangements adopted by HA in terms of risk to the client against the level of contractor's innovation input; and
- comparison of the strengths of different procurement modes in terms of key performance aspects: design quality, cost effectiveness, construction time efficiency, pre-contract lead time efficiency, innovation and construction quality control.

The benchmarking was made collectively among members of the working group based on their experience in the different procurement systems. The results are given in Figures 10.10 and 10.11. It shows that the working group considered that the integrated contract with three-envelope tendering system had enabled the highest extent of integration of contractor's innovation and contained within reasonable risk to the client in terms of certainty of cost, quality and time before committing to build.

In terms of the strengths of different procurement modes, the working group was content with the remarkable performance achievements of the integrated contract with three-envelope tendering system.

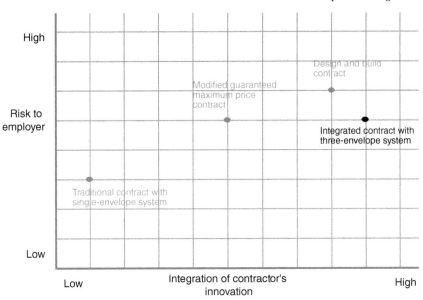

Figure 10.10 Benchmarking of risk vs innovation of different procurement modes

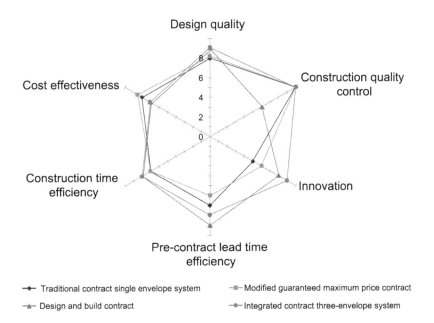

Figure 10.11 Benchmarking of the strengths of different procurement modes

Conclusion and future applications

The linear approach of the conventional design–bid–build procurement model has been working effectively with low risk and providing high reliability to the HA, but it is not conducive to integrating expertise of different contracting parties sufficiently early in the procurement process to make any significant innovative contributions and to deliver better value and quality products with improved productivity.

The innovative three-envelope tendering system for integrated contracts enables earlier and better integration of contractor's expertise with improved productivity and emphasises innovation in the delivery of public housing development. It was also successful in encouraging contractors' development and submission of design and innovation proposals and in catering for objective evaluation and selection of innovation proposals which were over and above the client's requirements at the tender evaluation stage.

With the successful and fruitful implementation in Kai Tak Site 1B and Anderson Road Sites 1 and 2 Phases A and B, the HA may consider adopting this tendering system again for other suitable mega-projects in the future. In order to gain maximum benefit from the three-envelope system, the system would be more appropriate for public housing developments with the following characteristics:

- a sizeable project which affords the opportunities for innovative design solutions with economies of scale. The project site can also provide ample works areas for applying advanced construction techniques on site, such as on-site pre-casting, etc;
- a standalone site within a new development area which entails low planning risk. Its interface with neighbouring sites should be simple and thus the client's requirements can be succinctly defined; and
- complex and large-scale housing projects as the contractor's expertise can be synergistically integrated upfront and innovations can be solicited and accommodated in the project to enhance productivity and the built quality, e.g. the presence of existing piles in Kai Tak Site 1B and the used quarry site in Anderson Road Site 1 and 2 Phases A and B, which provides an opportunity for innovative design to be aptly developed.

Following on from these two successful cases, the HA tried to apply this procurement approach for the third time in 2015 for a sizeable site in Queen's Hill producing over 12,000 flats, but in vain. Three deterrent factors were perceived by contractors in their feedback:

- Planning risk exists, and the scope of design and construction works was too complex for contractors to manage with ease. The project included public transport interchange and plenty of community facilities in addition to domestic, retail and car-parking facilities. The scope of work included demolition, site formation, roadworks and infrastructural works, as well as foundation and building works. The various consents and approvals required from statutory authorities in association with the complex development parameters were expected to impede innovations.

- The quantum of production was too large and beyond the financial viability of many experienced contractors in the industry, and, coupled with the HA's imposition of workload capping limits to avoid concentration risks, contractors considered that traditional procurement would be more favourable to both contracting parties. Projects with the development of 7,000–10,000 domestic flats are probably the optimum size for application of the three-envelope system, offering economies of scale and manageable by contractors in providing innovative design solutions.
- In 2015 Hong Kong faced an economic boom and a surge of workload for the construction industry and an acute shortage of human resources. The sheer volume of work during the tender stage for identification and exploration of innovations was too demanding on the project and construction teams when resources were fully stretched.

The lessons learned from Queen's Hill indicate that the three-envelope tendering system may not be suitable for all mega-projects, having regard to tenderers' perceived risks and market forces, economic and quantum risks, inherent planning and technical risks for specific sites. Hence, each mega-project must be considered on its own merits.

On reflection, the three-envelope system motivates innovation in the delivery of public housing development and steps forward to capitalise on the potential for synergistic integration of all contracting parties and professionals for the continuous and sustainable advancement for design and construction of the works. This brings about advancements in technology in the construction industry.

This has successfully transformed the tender stage into a 'learning programme' for all parties involved. These innovations will bear fruit in delivering cost-effective and quality housing and will benefits the public. It results in a win–win–win outcome for all, for projects of suitable size and complexity launched at the right time.

References

Abdul Rashid, R., Mat Taib, I., Wan Ahmad, W. B., Nasid, M. A., Wan Ali, W. N. and Mohd Zainordin, Z. (2006). Effect of procurement systems on the performance of construction projects. In: *International Conference on Construction Industry 2006: Toward Innovative Approach in Construction and Property Development, June 2006.* Available at: http://epublication.fab.utm.my/191/ (Accessed 11 August 2016).

Haugbølle, K. and Boyd, D. (2013). *Research Roapmap Report: Clients and Users in Construction.* CIB Publication 371. Rotterdam: International Council for Research and Innovation in Building and Construction (CIB).

Hong Kong Government (2009). *Gross Domestic Product. Fourth Quarter 2008.* Government of the Hong Kong Special Administrative Region, Census and

Statistics Department. Available at: www.censtatd.gov.hk/fd.jsp?file=B10300012008Q Q01B0100.pdf&product_id = B1030001&lang = 1 (Accessed 11 August 2016).

Hong Kong Housing Authority (2015). *Hong Kong Housing Authority Annual Report 2014/15*. Available at: www.housingauthority.gov.hk/mini-site/haar1415/en/index. html (Accessed 11 August 2016).

Koskela, L., Howell, G., Ballard, G., and Tommelein, I. (2002). *The Foundations of Lean Construction. Design and Construction: Building in Value.* Oxford: Butterworth-Heinemann.

Masterman, J. (2002). *Introduction to building procurement systems. Second edition.* London and New York: Spon Press.

11 BIM for clients

Developing digital dividends

Niraj Thurairajah and David Boyd

Introduction

Clients exist as enterprises within the wider environment of businesses and society as pointed out in the CIB W118 Research Roadmap (Haugbølle and Boyd, 2013). For a long time, businesses have used digital technologies as a way of improving performance and customer satisfaction. However, they have held construction and facility-related information separately and rarely used it to assist business processes. Clients can now see this information as an integral part of business processes to compete in this ever changing business world. Successful business and service providers need to start using technology to integrate this information as part of their product or service delivery. Thus, the use of Building Information Modelling (BIM) is the start of a move away from traditional practices and the formation of digital models to enable better decision making throughout the planning, design, construction and operation stages of a facility's life. Therefore, it is necessary for clients to appreciate and embrace the BIM way of working to move their business processes forward. However, clients should also understand the barriers, constraints and challenges associated with BIM and its implementation if they are to use this effectively within their business. While it is important to understand technological and process issues and constraints, people-related challenges are equally significant and need to be carefully managed to leverage potential benefits of BIM and wider digitisation. This requires intelligent clients with a clear purpose for their business development, and this chapter aims to explain how the wider adoption of digital tools, stimulated by BIM, will transform the business operation of buildings.

Developing perspectives of BIM

BIM is beginning to change the way buildings look, the ways in which they are designed and built and the way they function (Eastman *et al.*, 2011). There is a wealth of research on how BIM can be used for purposes such as a modelling tool, information tool, communication tool and facilities management tool (Popov *et al.*, 2006). Some authors focus on the technological elements (Smith and Tardiff, 2009; Lewis, 2012), which demonstrates the success of computer modelling, whereas others believe the main focus is the organisational transformation (Eastman *et al.*, 2011; FM and Beyond, 2011; O'Grady, 2013). The technological change of BIM is substantial and produces many efficiency benefits, but key to the success of BIM is a change in the way projects are undertaken and managed by the incorporation of a fully integrated design and project management process. The developing perspectives are shown in Figure 11.1, starting from seeing BIM just as a 3D modelling tool, which has recently been overtaken by the idea that it is a project management tool; however, the extension of this to BIM being a system provides the wider integrating advantages of digital information.

The narrowest and commonest understanding of BIM is first that it is a 3D model used for design and visualisation purposes. BIM as a concept has been around for a surprising length of time, though most academics agree that the first realisation of the concept, involving virtual buildings, was in the 1970s (Eastman *et al.*, 1974, 2011). In the 1970s CAD was introduced and it changed the design industry throughout the 1980s and 1990s. However, 2D CAD was only able to produce 2D drawings, and so, therefore, to reduce errors, technology has naturally evolved from 2D to 3D modelling (Yan and Damain, 2008), which has led to the introduction of BIM (Vogt, 2010). As a response to the increasing complexity and pressures of construction projects, BIM has been developed as an advanced technology to fulfil the current needs of users and to bridge the inefficiencies and fragmentation of older information and communication technology systems. Dimyadi *et al.* (2007) differentiate BIM and CAD in terms of design process whereas Eastman *et al.* (2008) considers the difference more holistically through comparing the BIM solid modelling benefits over CAD. What brought the concept of BIM to the forefront was a white paper aptly titled 'Building Information Modelling' by software vendor Autodesk in 2003 (who also introduced the acronym 'BIM').

However, BIM not only accesses design information in 3D, but also provides the capability to access information across a fourth dimension (time), fifth

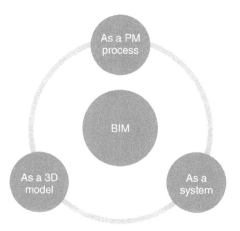

Figure 11.1 Perspectives on BIM

dimension (cost) and others. BIM also has a key role in building maintenance, as the model information can be used across a building's entire lifecycle. Indeed, given that the operation and maintenance of a building equates up to 60 per cent of the overall costs of a facility (Akcamete *et al.*, 2011), BIM is potentially most valuable to facilities managers, allowing information about the use of the building to be collated and analysed, and so enabling the operating costs to be effectively controlled. BIM as a 3D model, then, is a term used for defining the process of information management through the life cycle of a building.

Secondly, an increasing number of industry professionals and clients are starting to see BIM 'as a Project Management (PM) process' that improves planning and collaboration during building design and construction. Here BIM not only allows a single federated building model to be created that integrates information supplied from all disciplines involved in the project team (Succar, 2009), but also enhances collaboration in the design and the construction stages. BIM has the potential to promote further integration between parties and enable better flow of information across the whole building process especially the operational stage. It is clear then that BIM is far more than simply a move to computer 3D modelling and in reality represents a fundamental change in the way projects are undertaken and managed and so will result in essential changes to the business process and workflows (Smith and Tardiff, 2008).

Thirdly, and less well understood, is 'BIM as a system' which provides an opportunity to use building information for business intelligence that could yield numerous benefits. The wider system understanding of BIM, then, is where the potential for clients and users are unlocked. However, such a system view requires BIM technology and the federated model in its core to gather operational information for business intelligence. Furthermore, this system needs to work alongside other business intelligence systems used within the client organisation. The creation of a whole enterprise system (Giachetti, 2010) based on information collection and transformation becomes the ultimate achievement from the adoption of BIM.

BIM implementation: state of the world

The property and construction industry has many challenges which derive from the complex social, economic and environmental conditions of the built environment. This means that BIM will be constituted differently in different parts of the world. Much of this can be related to the way governments want to control development and enterprise or whether they want to let this be determined through a market system that operates independently of government. However, in most countries the drive for BIM has come from governments seeing the advantage of BIM to themselves as a long-term asset holder and as an economic stimulus for their industry.

McGraw Hill (2014) has investigated the evolution of BIM since 2007 through extensive global surveys and concluded that BIM has dramatically changed during the past few years. Moreover, the survey results show varying levels of BIM adoption in the United States, the United Kingdom, France, Germany and Canada compared to Australia, Brazil, Japan, Korea and New Zealand. Increasing use of BIM is seen in Europe due to the decision of the European Parliament to change European public procurement rules to implement electronics tools such as BIM (Smith, 2014). Furthermore, Smith (2014) claims countries have different initiatives and approaches for BIM implementation. Implementation and understanding of BIM is fundamentally gained through guidelines, strategies, standards and protocols. This shows that governments' involvement and BIM guidelines and standards play an important part in adopting BIM. In addition to skills shortage, willingness of institutions to embrace such an open, collaborative and digital tool is also crucial to reaping the benefits of digital transformation.

USA

The US construction industry has committed to using BIM since 2003 through strategic and incremental adoption of 3D, 4D and BIM technologies. Wong *et al.* (2009) claimed the US to be one of the global leaders in BIM implementation in construction industry. The General Services Administration (GSA) which is responsible for construction and operation of all federal facilities in the US has initiated BIM on public projects since 2006. GSA introduced several guidelines and standards including a National BIM standard which is accepted internationally. GSA plans to introduce BIM throughout the whole lifecycle including using spatial programming validation, 4D phasing, laser scanning, energy and sustainability, circulation and security validation and maintenance of building elements (BuildingSMART Australasia, 2012).

Europe

In Europe, the Finnish government, one of the early adopters, has produced a full series of implementation guides based on feedback from many BIM pilot studies. Similarly, in Norway the government building agency, Statsbygg, decided to implement BIM for the whole life cycle of their facilities. BIM was slowly adopted from 2007 and by 2010 Statsbygg projects were using IFC/IFD-based BIM. Interestingly, the Norwegian Homebuilders Association has encouraged the industry to adopt BIM and IFC. The Danish government also encourages BIM implementation and has invested heavily in BIM-related research and development (Smith, 2014). Danish government clients such as The Palaces and Properties Agency, the Danish University Property Agency and the Defence Construction Services have agreed to adopt BIM in their projects (BCA, 2012). BIM has been used in Danish construction projects since 2007; however, IFC-compliant BIM was required from 2010 (BuildingSMART Australasia, 2012).

Germany is currently struggling to adopt BIM in their construction projects due to the legally protected professional titles and defined work and fee scales. Germany's Digital Building Platform, a BIM task group set by trade associations has laid the groundwork for public sector BIM adoption including standardising of process and device description, and developing guidelines for digital planning methods with the provision of sample contracts. In France, the Ministry of Housing (Ministère du Logement) has recently decided to implement BIM in their housing and general construction. It has been announced that 500,000 houses in France will be developed using BIM by 2017.

The UK government mandated the use of level 2 BIM in governmental construction projects from 2016 (Cabinet Office, 2011). Level 2 BIM means each party working on their own model but collaborating using a federated model. This and library management is targeted to save 20 per cent in the procurement cost and 50 per cent in carbon emissions (Cabinet Office, 2011). The BIM Task Group was established as part of this movement to monitor both public sector clients and private sector supply chain in re-engineering their work practices to facilitate BIM delivery (McGraw Hill, 2014). British Standard BS1192 (Richards, 2010) and its associated parts are the primary BIM standards in the UK.

Australasia

There is no mandatary use of BIM in Australian construction projects; however, during the past five years various parties have shown interest in adopting BIM in their construction projects due to its benefits (CIBER, 2012). This has led to the creation of Australian BIM guides such as the 'National BIM Guide' by the National Specifications (NATSPEC), 'National Guide for Digital Modelling' by the Corporate Research Centre for Construction Innovation (CRC-CI), the 'Australian and New Zealand Revit Standards' (ANZRS) and the BIM-MEPAUS guidelines and models.

The Singapore Building and Construction Authority (BCA, 2012) in 2008 introduced the first BIM electronic submission in the world (e-submission). Following that in 2010 it established a roadmap to implement BIM in all construction public projects by 2015 (Smith, 2014). The aim of this was to encourage 80 per cent of the construction projects to adopt BIM and to improve productivity by 25 per cent over the following decade. The Singapore government has also established a Construction Productivity and Capability Fund (CPCF) of EUR 15.8 million with BIM as a key target. The Construction and Real Estate Network (CORENET) program was introduced in 2000 as a strategic initiative to change the industry with the aid of information technology (Smith, 2014). CORENET enables exchanging the information among all the project participants and the CORENET e-plan checks system, 'Integrated plan checking', allows architects and engineers to check their BIM-designed buildings for regulatory compliance through an online 'gateway'. This has increased the number of public–private initiatives using BIM in their large pilot projects. Singapore has also adopted Industry Foundation Classes (IFC) standards during BIM implementation (BuildingSmart Australasia, 2012).

Similarly, Hong Kong has slowly started using BIM: Hong Kong Housing Authority has required BIM for all new projects since 2014. Japan is becoming active in BIM engagement. Construction companies in Japan have started using

BIM mostly for supply chain management, model-driven robotics and post-construction activities (Smith, 2014). Moreover, Japanese companies have reported a positive Return on Investment (RoI) with BIM implementation.

In an analysis of the factors affecting the success of BIM adoption in different countries, Fenby-Taylor *et al.* (2016) developed an 'Ease of Integration Index' which they relate to the 'Gross Fixed Capital Formation'. The latter is significant as it is the measure of the creation of fixed capital, which is the best economic indicator of activity in the built environment. Their Ease of Integration Index compounds data on economics, policy and governance capabilities in each country. In their analysis they put Singapore and Norway in the lead of development as both countries have high Gross Fixed Capital Formation, which makes the use of BIM economically worthwhile whilst at the same time both have strong governance and policy capabilities driven by their governments. They recognise that this analysis promotes the role of large businesses and so make a key recommendation that policy on BIM development needs to explicitly address small and medium-sized enterprises as these are the loci of entrepreneurship and innovation.

Benefits of using BIM

Table 11.1 lists the benefits of using BIM against the three perspectives of BIM in Figure 11.1 and presents the authors promoting these. Under the category of 'BIM as a model', it is evident that visualisation is seen as the key benefit (Azhar *et al.*, 2012; Arayici *et al.*, 2011; Li *et al.*, 2008). Visualisation not only allows designers to better understand what they are creating but also allows the designs to be interrogated and communicated more effectively by others. Also in this category, clash detection is a powerful tool to eliminate constructability problems where these had been previously rectified expensively on site (Bryde *et al.*, 2013; Cao *et al.*, 2015). Studies from other industries suggest that use of parametric modelling will increase and even move to the next generation of generative modelling. These features involve greater automation of design within the industry but at the same time provide a new opportunity to undertake design differently (Kensek and Noble, 2014). In this the system will automatically produce multiple designs which address the initial constraints that have been determined and coded into the system; selection and advancement of one of these solutions is then carried forward by human intervention. This idea of meta-modelling, that is, modelling of the model, can be applied in any area constituted in information terms.

Under the category of 'BIM as a project management process', the key benefit is effective collaboration and communications (Arayici *et al.*, 2011; Kymmell, 2008; BSI, 2010). In the complex interdependent world that is construction, where many parties have to work together to produce the finished building, collaboration and communications are key to smooth processes and the avoidance of problems. A further benefit in this category is the planning and scheduling of activities where the sequencing of construction can be reviewed and different

Table 11.1 Potential benefits of BIM

Potential benefits of using BIM	Grilo and Jardim-Gonçalves (2010)	Azhar (2011)	Eastman et al. (2011)	Sebastian (2011)	Jung and Joo (2011)	Ku and Taiebot (2011)	Arayici et al. (2011)	Deutsch (2011)	Schade et al. (2011)	Azhar et al. (2012)	Barlish and Sullivan (2012)	Nawari (2012)	Bryde et al. (2013)	Eadie (2013)	Love et al. (2014)	Li et al. (2008)	Latiff et al. (2015)	Cao et al. (2015)	Matthews (2015)
As a modelling and visualisation tool																			
Better design via design validation	x						x	x		x	x					x		x	
Clash detection		x	x	x			x	x			x		x				x	x	x
Quick solutions for design issues			x																
Quicker understanding of design change orders			x							x	x		x			x			
Quicker simulations										x	x						x		
Improved visualisation	x			x	x					x	x		x	x		x	x	x	
As a project management process																			
Effective collaboration and clear communication	x	x	x	x	x		x			x	x			x		x	x		
Faster and effective processes		x																	
Better decision making	x				x		x			x	x		x			x			x
Increased productivity and profitability						x												x	x
Time management												x			x	x		x	
Reduced errors and risks																		x	
Reduced waste/lean approach							x							x		x	x		

Table 11.1 continued

Potential benefits of using BIM

	Grilo and Jardim-Goncalves (2010)	Azhar (2011)	Eastman et al. (2011)	Sebastian (2011)	Jung and Joo (2011)	Ku and Taiebot (2011)	Arayici et al. (2011)	Deutsch (2011)	Schade et al. (2011)	Azhar et al. (2012)	Barlish and Sullivan (2012)	Nawari (2012)	Bryde et al. (2013)	Eadie (2013)	Love et al. (2014)	Li et al. (2008)	Latiff et al. (2015)	Cao et al. (2015)	Mathews (2015)
Understanding the project and monitoring	x												x						x
Creating sustainable communities														x					
Early supply chain involvement					x							x							
Organised project schedule and budget										x			x	x	x		x	x	
High quality project outcomes						x					x		x		x				x
Multidisciplinary coordination																x			
Integrated design with cost estimation database			x	x				x		x								x	
As an information system																			
Better value through better understanding of lifecycle costing	x		x	x		x				x	x		x		x		x		x
Improved data integrity and efficiency											x		x						
Distributed access and retrieval of building data															x			x	
Better data documentation through clarified and clear information															x		x		
Quick data preparation through automated data assembly	x				x			x											
Reuse of requirements and project information in facility management	x															x			
Rigorous analysis of building performance	x								x										
Predictable environmental performance	x								x										

methods assessed (Fischer and Kunz, 2004; Haymaker *et al.*, 2005). The use of visualisation of the project process can further enhance the collaboration benefits.

As regards the benefits of 'BIM as an information system', then, it is the ability to better predict life cycle costs (Eastman *et al.*, 2011; Sebastian, 2011) and to simulate environmental performance (Schade *et al.*, 2011; Azhar *et al.*, 2012) that are key at the moment. The idea that clients could specify and manage building performance which would enable the occupying organisations to perform better is a growing aspiration (Mayouf *et al.*, 2014). Although, this idea has been around for many years, it is only realisable now that BIM can be implemented so that building management information can be collected then drawn together to be used as the focus of building development.

Challenges and constraints of using BIM

Even with much more technical maturity and greater experience, BIM will not overcome all of the challenges in the property and construction industry. In addition, the use of such a sophisticated tool produces its own problems and requires its own expertise. This does not mean that BIM cannot add value or will disappear from the construction industry, but these issues need to be taken into account and managed through BIM use and development. Table 11.2 presents the key challenges and constraints for implementing BIM in construction industry together with the authors who highlight these. The table is divided into five sections: Implementation; Technical issues; Social/organisational issues; Legal/ contractual issues; and Learning issues. These are not independent but identify the domains that are important to manage to develop BIM; importantly they will occur together and this requires the management to be more sophisticated to overcome the problems.

Implementation is a major problem of all new technologies and must be recognised as a real cost; just purchasing software and software training will not mean that it can be used effectively. One of the key aspects is the way that BIM needs to be made to work with current practices (Grilo and Jardim-Goncalves, 2010). In many ways BIM requires a major change to the way design and construction is undertaken and processes and procedures need to be completely rewritten.

The social and organisational aspects then become more important as these can drive the development including positively overcoming problems or if not addressed can become the site of failure (Gu and London, 2010). The fundamental change in social environments can be disruptive (Azhar, 2011) and sophisticated analysis is required to understand outcomes (Harty, 2008). These organisational problems are not confined to operational matters but challenge management decision making and strategies (Dossick and Neff, 2010).

Given the promotion of the technical capabilities of BIM, it is surprising that there are still major technical issues. These include software operational problems (Jung and Joo, 2011; Ku and Taiebat, 2011) which will be overcome as software develops but be repeated with new features; however, more complex are

Table 11.2 Challenges and constraints of using BIM

Challenges and constraints of using BIM	Grilo and Jardim-Goncalves (2010)	Gu and London (2010)	Jung and Joo (2011)	Sebastian (2011)	Azhar (2011)	Ku and Taiebat (2011)	Azhar et al. (2012)	Bartish and Sullivan (2012)	Goucher and Thurairajah (2012)	Bryde et al. (2013)	Alabdulqader et al. (2013)	Kunal et al. (2013)	Stanley and Thurrel (2014)	Smith (2014)	Tulenheimo (2015)	Latiff et al. (2015)	Harry (2008)	Dossick and Neff (2010)
Implementation																		
Integrating BIM with current practices	x																	
Investment cost			x							x		x	x			x		
No single set of implementation guidelines					x	x	x						x					
Insufficient practice strategies for integration					x													
Social and organisational change																		
Fundamental change		x								x	x	x	x	x				x
Uncertainty over control		x	x	x										x			x	x
Resistance to change			x			x				x	x	x	x		x		x	x
Threat to viability of profession/roles									x						x	x	x	x
Time taken to use the model					x						x				x	x		
No suitable procedures								x										x

Technical issues

Software problems			x		x			x	x
Lack of hardware support	x							x	x	x
Different projects and company requirements	x		x						
Interoperability challenges		x	x	x			x	x	
Incorporating unique items		x							

Legal/contractual issues

Lack clarity of ownership of BIM data		x				x		x	
Protection from copyright laws and legal channels		x		x		x			
Licensing issues		x							x
Limited warranties and disclaimers of liability by designers		x		x					
Lack of policies/standards/protocols		x	x	x		x	x		
Insurance issues			x		x	x	x		

Learning and experience

Varying levels and lack of team knowledge	x		x						
Lack of client demand/interest				x					
Lack of training/education/learning curve	x	x	x		x	x		x	x
Lack of clarity of roles and responsibilities	x		x	x		x	x	x	

interoperability issues (Azhar, 2011) where it is difficult for different software to share information reliably. The use of IFCs in this respect may help the interoperability problem but few software packages have this as their primary feature.

Many authors (Singh *et al.*, 2011; McAdam, 2010) highlight the legal and contractual issues as being the most significant problems of full BIM implementation. First, the risk and liability of using new software create many legal issues. While different project participants access the project information the reliability and confidentially of this information (Li *et al.*, 2008) cannot be guaranteed. Secondly, due to the fact that models integrate different pieces of information, contributed by different members of the multidisciplinary project team, ownership of problems cannot be vested in a particular party (Sebastian, 2011).

Finally, the industry's reluctance to learn new methods has been identified as a major barrier to BIM implementation (Johnson and Laepple, 2003). This is compounded by the fact that the cost of training and resources is high (Azhar *et al.*, 2011; Crotty, 2012; Yan and Damian, 2008). As a result, Ku and Taiebat (2011) indicate that the industry uses self-learning, seminars and workshops, in-house training and hiring previously trained personnel to overcome these issues. This approach may reproduce problems rather than tackling them head on.

These barriers and constraints become more significant as we rely more and more on the model to undertake work automatically. This also creates further social, economic and political tensions which will intervene in the pure technological development of BIM. A further problem is that the adoption of BIM is an opportunity for the promotion and development of self-interest both of individuals in projects and by companies in the support of the industry who wish to lock-in organisations to their software. As such there will be winners and losers and this may produce difficult decisions for clients as to which side to support and the one that appears to deliver most may not be the best long-term solution.

Approach to digital dividends

It is most important for clients to realise that a digital transformation is happening around them and the most successful businesses of the future are likely to be those that maximise their current investment to exploit these undeveloped possibilities. Digital technologies such as BIM can bring more choice and greater convenience in using building spaces. It provides an information system with a shared facility model that can be the backbone of operations, maintenance and scenario planning in the whole life span of the portfolio and of their facilities. However, the broader development benefits from using BIM have not been realised. The challenging parts of asset lifecycle management such as early involvement in the design and handover of building information are still challenging within BIM projects. To secure digital dividends from BIM, development strategies need to be broader than IT strategies. Clients need to create complementary conditions for BIM to be effective. Hence clients need a strong awareness of how all aspects of their business can work with digital information, and how this can be integrated from different systems whilst at the same time understanding analogue information and its social and economic value. It is in this that BIM can be truly transformational. Figure 11.2 shows how the digital dividend can grow out of BIM by focusing on three areas: accessibility, efficiency and transformation.

Accessibility involves the ability to easily search, identify and visualise related information. This requires not only understanding the sources of information, that is, what can be measured and calculated, but more importantly its characteristics and meaning within the differently roles of the organisation. For example, BIM provides representation of actual buildings components as BIM objects. These BIM objects have metadata such as type, size, geometry, classification, expected performance and material that describe it. These objects can be generic or specific and component or material. They are also the carriers of history, both about the individual object and for the asset as a whole. Information behind these objects can be searched and visualised for design coordination, construction document production, scheduling, maintenance and operation. Unfortunately, the current practice is to leave generic objects without specific information for as-built and O&M (Operation and Maintenance) documentation, which hinders life-cycle management of facilities. To secure the dividends from BIM, accessibility of specific information from manufacturers and contractors is essential. Moving on to the business, accessibility refers to the use of space and the business processes that take place within. These need to be managed not just as separate tasks but as integrated data environments. In collecting and managing this data, future opportunities can be identified for new uses of space and new processes that work within them.

Efficiency, on the other hand, comes from automation and coordination of activities. For example, parametric modelling aspect of BIM improves efficiency and supports benefits such as clash detection, automatic quantity take-off and multidisciplinary coordination. In this regard we are already starting to see online toolkits that can define, manage and validate responsibility for information development and delivery at each stage of the asset lifecycle (NBS, 2016). However, clients don't directly benefit from these operational efficiencies, although building costs may go down as a result of more efficient and less repeat work. However, it is operational and maintenance efficiencies that are the triumphs of BIM for clients. A rich handover model with linked documentation

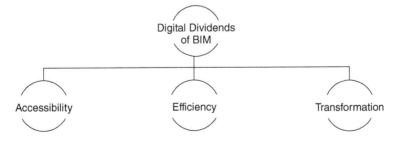

Figure 11.2 Digital dividends of BIM

that includes the parameters needed for operational management can provide clients with a new degree and efficiency of management of the operation of the building. The federated as-built model connected to in-built devices that capture live field information provide a new active management and learning feedback about the building in use. The model can also receive data from its individual components with geo-positioning information, RFID (Radio Frequency Identification Devices) and NFC (Near Field Communication) tags that can assist with maintenance and reconfiguration. This efficiency dividend is already familiar to businesses that have been using information for a long time to improve their efficiency in the form of KPIs and management accounting measures. Thus, the idea about BIM, associated with information about the building space, provides a new opportunity both at the process level and, just as importantly, in creating environments for operational efficiency. This leads to the third area for the digital dividend.

Transformation is the most important digital dividend of BIM and a digitally driven client organisation. This transformation envisages buildings as part of a smart, networked world enabling a new digital economy and new business models. As pointed out in *Digital Built Britain* (HM Government, 2015: 14–15), BIM and other digital technologies have the:

> ability to bring together through open data standards from design, construction and operations and across market sectors offering the ability to analyse and create the learning feedback loops that industry needs to be able to deliver sustainable long term improvements in asset performance.

BIM, then, is part of the transformation of the organisation, and even the driver behind creating a data-driven business supported by a business information system. The digitisation of management information systems allows businesses to operate in new ways and to become digital enterprises (Laudon and Laudon, 2014). Business operations is the most obvious addition where the business or service is seen as a machine and, if not automated, then uses information technology to interacted with subjects and to integrate the divisions of the enterprise. Such systems have been available for a number of years and are often sold as ERP (Enterprise Resource Planning) software packages. These provide a platform around which business operates, connecting finance, human resources, product planning, purchase, manufacturing or service delivery, marketing and sales, inventory management, shipping and payment. ERP systems have now been developed beyond the back office operations to encompass the front office (Laudon and Laudon, 2014). Here the opportunities expand enormously with Customer Relationship Management (CRM) and spread into e-business systems and supply chain management. Other opportunities exist that would be useful for public services; here approaches, such as e-government, can allow direct and immediate relationships with the public and could be used for real time democratic decision making.

However, digitisation also offers new opportunities to see the enterprise differently by collecting data from across the business and from its environment

(Peitz and Waldfogel, 2012). This is often referred to as Big Data, which allows the handling of both structured and unstructured data. This is a growing facility and there are still problems of effectively handling unstructured data and for ensuring security. This field is developing rapidly. All equipment will eventually have inbuilt sensors which send data back about short-term and, more importantly, long-term operation such that the enterprise can learn from its history. This also includes the operation and usage of buildings so that we can learn in a new way how buildings perform better. To make this happen, then, systems need to handle both semi-structured and unstructured data. This requires systems that undertake searches and understand context so that there is an awareness of meaning through metadata. This involves qualitative and quantitative techniques of extracting and categorising data to analyse behavioural data, trends and patterns across the business (Correa, 2015).

This availability of performance data on a wide scale can enable smarter decision making and new business models. *Digital Built Britain* (HM Government, 2015) promotes further nationwide use of this data for optimal maintenance and operations schemes, investment opportunities and to deliver key services. Intelligent clients would see their business as part of the new digital economy and form new business relationships and innovative platforms to provide effective and efficient service provision. An example is the transformation of the housing industry, as promoted by Boyd *et al.* (2015), where the industry governs itself around a new integrated and long-term agenda based on information. This benefits not just clients, whether developers or users, but also the industry in its efficiency and long-term engagement, the environment and its governance through transparent and interactive processes, and the community in terms of skills, jobs and economic return. For example, building maintenance becomes a new proactive business, led by big data on operations, and providing a long-term role for the construction industry which benefits users.

Conclusion

Although BIM is the most obvious and promoted application of digital revolution for clients, it is only the starting point of a potentially momentous and profitable transformation of their complete business or service provision. This digital dividend needs to be invested in, in order to take advantages as opportunities come along. This means clients should invest now in getting experience and gaining insight into these opportunities, otherwise they will always be behind in development. They will suffer a disjointed adoption in the future with the potential that a new economic world forces out their business before it has a chance to transform. Equally, on the public sector side, there are new opportunities in this transformation for developing democracy, inclusion and community engagement as well as regional and enterprise development. This promotes the potential of clients to drive innovation, not just in construction, but also in digital business, for their mutual benefit. There are dangers particularly of data security and software lock-in. However, the opportunities to avoid these require businesses and services to be working on this now and proactively determining how the transformation will take place rather than just waiting to be the recipient of it. The problems in different locations around the world will be different as the governance and digital infrastructure are different and in different stages of maturity. There will be globalisation trends which will challenge national differences and put pressure on individual countries to adopt international approaches; such tensions need to be acknowledged and managed. This is all happening now; there is really no other position than for clients to get involved.

References

Akcamete, A., Lui, X., Akinci, B. and Garret, J. H. (2011). Integration and visualization maintenance and repair work orders in BIM: Lessons learned from a prototype. *Proceedings of the 11th International Conference on Construction Applications of Virtual Reality (CONVR)*. Weimar, 3–4 November.

Arayici, Y., Coates, P., Koskela, L., Kagioglou, M., Usher, C. and O'Reilly, K. (2011). BIM adoption and implementation for architectural practices. *Structural Survey*, Vol. 29 (1), 7–25.

Azhar, S. (2011). Building information modeling (BIM): Trends, benefits, risks, and challenges for the AEC industry. *Leadership and Management in Engineering*, Vol. 11 (3), 241–52.

Azhar, S., Khalfan, M. and Maqsood, T. (2012). Building information modeling (BIM): Now and beyond. *Australasian Journal of Construction Economics and Building*, Vol. 12 (4), 15–28.

Barlish, K. and Sullivan, K. (2012). How to measure the benefits of BIM: A case study approach. *Automation in Construction*, Vol. 24, 149–59.

BCA (2012). *Building Smart: The BIM Issue*. Singapore: Building and Construction Authority. Available at: www.bca.gov.sg/publications/BuildSmart/others/buildsmart_11issue9.pdf (Accessed 19 August 2016).

Boyd, D., Thurairajah, N. and Leonard, M. (2015). *Housing: The digital revolution*. Technical Report. Birmingham: Birmingham City University and the Building Alliance.

Bryde, D., Broquetas, M. and Volm, J. M. (2013). The project benefits of building information modelling (BIM). *International Journal of Project Management*, Vol. 31 (7), 971–80.

BSI (2010). *Constructing the Business Case: Building Information Modelling*. London and Kenley, Surrey: British Standards Institution and BuildingSMART UK.

BuildingSMART Australasia (2012). *National Building Information Modelling Initiative, Vol. 1 : Strategy for the focussed adoption of building information modelling and related digital technologies and processes for the Australian built environment sector*. Sydney: Department of Industry, Innovation, Science, Research and Tertiary Education.

Cabinet Office (2011). *Government Construction Strategy*. London: TSO.

Cao, D., Wang, G., Li, H., Skitmore, M., Huang, T. and Zhang, W. (2015). Practices and effectiveness of building information modelling in construction projects in China. *Automation in Construction*, Vol. 49, 113–22.

CIBER (2012). *Building Information Modelling (BIM): An Introduction and International Perspectives, Centre for Interdisciplinary Built Environment Research*. Research Report. Newcastle, NSW: The University of Newcastle.

Correa, F.R. (2015). Is BIM big enough to take advantage of big data analytics? *2015 Proceedings of the 32nd International Symposium on Automation and Robotics in Construction (ISARC)*, Vol. 32, 1–8.

Crotty, R. (2012). *Impact of BIM on the construction industry*. London: Spon Press.

Deutsch, R. (2011). *BIM and Integrated Design: Strategies for Architectural Practice*. Hoboken, New Jersey: John Wiley & Sons, Inc.

Dimyadi, J. A. W., Spearpoint, M. J. and Amor, R. (2007). Generating fire dynamics simulator geometrical input using an IFC-based building information model. *Journal of Information Technology in Construction*, Vol. 12, 443–57.

Dossick, C. S. and Neff, G. (2010). Organizational divisions in BIM-enabled commercial construction. *Journal of Construction Engineering and Management*, Vol. 136 (4), 459–67.

Eadie, R., Browne, M., Odeyinka, H., McKeown, C. and McNiff, S. (2013). BIM implementation throughout the UK construction project lifecycle: An analysis. *Automation in Construction*, Vol. 36, 145–51.

Eastman, C., Eastman, C.M., Teicholz, P., Sacks, R. and Liston, K. (2011). *BIM Handbook: A Guide to Building Information Modeling for Owners, Managers, Designers, Engineers and Contractors* (2nd edn). Hoboken, New Jersey: John Wiley & Sons, Inc.

Eastman, C., Fisher, D., Lafue, G., Lividini, J., Stoker, D. and Yessios, C. (1974). An outline of the building description system. *Research Rep*, Vol. 50, 2–23.

Eastman, C., Teicholz, P., Sacks, R. and Liston, K. (2008). *BIM Handbook: A Guide to Building Information Modeling for Owners, Managers, Designers, Engineers and Contractors*. New York: John Wiley & Sons, Inc.

Fenby-Taylor, H., Thompson, N., Philp, D., MacLaren, A., Rossiter, D. and Bartley, T. (2016). *Scotland Global BIM Study*. dotBuiltEnvironment. Scotland. DOI: 10.13140/RG.2.1.5108.8889

Fischer, M. and Kunz, J. (2004). *The Scope and Role of Information Technology in Construction*. Available at: http://cife.stanford.edu/online.publications/TR156.pdf (Accessed 19 August 2016).

FM and Beyond (2011). *BIM and IPD Making Value Engineering Irrelevant*. Available at: http://fmandbeyond.blogspot.co.uk/2011/02/bim-and-ipd-making-value-engineering.html (Accessed 19 August 2016).

Giachetti, R. (2010). *Design of Enterprise Systems: Theory, Architecture, and Methods*. Boca Raton, Florida: CRC Press.

Goucher, D. and Thurairajah, N. (2012). Usability and impact of BIM on early estimation practices: Cost consultants perspectives. In: *Proceedings of CIB Management of Construction: Research to Practice*, Vol. 2, 555–69. Available at: www.irbnet.de/daten/iconda/CIB_DC25669.pdf (Accessed 8 October 2016).

Grilo, A. and Jardim-Goncalves, R. (2010). Value proposition on interoperability of BIM and collaborative working environments. *Automation in Construction*, Vol. 19 (5), 522–30.

Gu, N. and London, K. (2010). Understanding and facilitating BIM adoption in the AEC industry. *Automation in Construction*, Vol. 19 (8), 988–99.

Harty, C. (2008). Implementing innovation in construction: Contexts, relative boundedness and actor-network theory. *Construction Management and Economics*, Vol. 26 (10), 1029–41.

Haugbølle, K. and Boyd, D. (2013). *Clients and Users in Construction. Research Roadmap Report*. CIB Publication 371. Rotterdam: CIB General Secretariat. Available at: http://site.cibworld.nl/dl/publications/pub_371.pdf (Accessed 28 August 2016).

Haymaker, J., Kam, C., and Fischer, M. (2005, July). A methodology to plan, communicate and control multidisciplinary design processes. In R. Scherer, P. Katranuschkov and S.-E. Schapke (Eds.). *CIB W78 22nd Conference on Information Technology in Construction, Dresden, Germany*. Dresden: Institute for Construction Informatics, Technische Universität Dresden.

HM Government (2015). *Digital Built Britain*. Available at: www.gov.uk/government/uploads/system/uploads/attachment_data/file/410096/bis-15-155-digital-built-britain-level-3-strategy.pdf (Accessed 19 August 2016).

Johnson, R. E. and Laepple, E. S. (2003). Digital innovation and organizational change in design practice. CRC Working Paper no. 2. CRS Centre, Texas A & M University.

Jung, Y. and Joo, M. (2011). Building information modelling (BIM) framework for practical implementation. *Automation in Construction*, Vol. 20 (2), 126–33.

Kensek, K, and Noble, D. (eds.) (2014). *Building Information Modeling: BIM in Current and Future Practice*. Hoboken, NJ: John Wiley & Sons.

Ku, K. and Taiebat, M. (2011) BIM experiences and expectations: The constructor's perspective. *International Journal of Construction Education and Research*. Vol. 7, 175–97.

Kymmell, W. (2008), *Building Information Modeling: Planning and Managing Construction Projects with 4D CAD and Simulation*. New York: McGraw-Hill Construction.

Latiffi, A. A., Brahim, J., Mohd, S. and Fathi, M. S. (2014). The Malaysian government's initiative in using building information modeling (BIM) in construction projects. In: Chantawarangul, K., Suanpaga, W., Yazdani, S., Vimonsatit, V. and Singh, A. (eds.) *The Second Australasia and South-East Asia Structural Engineering and Construction Conference* (ASEA-SEC-2 2014), 3–7 November 2014, Bangkok, Thailand. Fargo, ND: ISEC Press, 767–72.

Laudon, K. C. and Laudon, J. P. (2014). *Management Information Systems: Managing the Digital Firm* (13th edn). Upper Saddle River, NJ: Pearson Education.

Lewis, S. (2012). BIM: Who does what and when? Available at: www.building.co.uk/bim-who-does-what-and-when?/5044400.article (Accessed 19 August 2016).

Li, H., Huang, T., Kong, C. W., Guo, H.L., Baldwin, A., Chan, N. and Wong, J. (2008). Integrating design and construction through virtual prototyping. *Automation in Construction*, Vol. 17 (8), 915–22.

Li, N., Becerik-Gerber, B., Krishnamachari, B. and Soibelman, L. (2014). A BIM centred indoor localization algorithm to support building fire emergency response operations. *Automation in Construction*, Vol. 42, 78–89.

Love, P. E., Matthews, J., Simpson, I., Hill, A. and Olatunji, O. A. (2014). A benefits realization management building information modeling framework for asset owners. *Automation in Construction*, Vol. 37, 1–10.

McAdam, B. (2010). Building information modelling: The UK legal context. *International Journal of Law in the Built Environment*, Vol. 2 (3), 246–59.

McGraw Hill (2014). *The Business Value of BIM for Construction in Global Markets*. Bedford, MA: McGraw Hill Construction.

Matthews, J., Love, P. E., Heinemann, S., Chandler, R., Rumsey, C. and Olatunj, O. (2015). Real time progress management: Re-engineering processes for cloud-based BIM in construction. *Automation in Construction*, 58, 38–47.

Mayouf, M., Boyd, D. and Cox, S. (2014). Perceiving space from multiple perspectives for buildings using BIM. In: Raiden, A. B. and Aboagye-Nimo, E. (Eds). *Proceedings of 30th Annual ARCOM Conference, 1–3 September 2014, Portsmouth, UK, Association of Researchers in Construction Management*, 683–92.

Nawari, N. O. (2012). BIM standard in off-site construction. *Journal of Architectural Engineering*, Vol. 18 (2), 107–13.

NBS (2016). *Periodic Table of BIM: Enabling Tools*. Available at: www.thenbs.com/knowledge/periodic-table-of-bim-enabling-tools (Accessed 19 August 2016).

O'Grady, M. (2013). *Improving BIM Outcomes through Industry Communication, Collaboration and Consolidation*. Available at: www.bim-in-practice.com.au/Event.aspx?id=825586 (Accessed 12 April 2013).

Peitz, M. and Waldfogel, J. (Eds.) (2012). *The Oxford Handbook of the Digital Economy*. Oxford: Oxford University Press.

Popov, V., Mikalauskas, S., Migilinskas, D. and Vainiūnas, P. (2006). Complex usage of 4D information modelling concept for building design, estimation, scheduling and determination of effective variant. *Technological and economic development of economy*, Vol. 12 (2), 91–8.

Richards, M. (2010). *Building Information Management: A Standard Framework and Guide to BS 1192*. London: BSI Standards.

Schade, J., Olofsson, T. and Schreyer, M. (2011). Decision-making in a model-based design process. *Construction management and Economics*, Vol. 29(4), 371–82.

Sebastian, R. (2011). Changing roles of the clients, architects and contractors through BIM. *Engineering, Construction and Architectural Management*. Vol. 18 (2), 176–87.

Singh, V., Gu, N. and Wang, X. (2011). A theoretical framework of a BIM-based multi-disciplinary collaboration platform. *Automation in Construction*, Vol. 20 (2), 134–44.

Smith, D. K. and Tardif, M. (2009). *Building information modeling: A strategic implementation guide for architects, engineers, constructors, and real estate asset managers*. Hoboken, NJ: John Wiley & Sons, Inc.

Smith, P. (2014). BIM implementation: Global strategies. *Procedia – Engineering*, Vol. 85, 482–92.

Stanley, R. and Thurnell, D. (2013). Current and anticipated future impacts of BIM on cost modelling in Auckland. In: Yiu, T. W. and Gonzalez, V. (Eds.). *Proceedings of 38th AUBEA International Conference, 20–22nd November 2013, Auckland, New Zealand*. Auckland: University of Auckland. Available at: www.library.auckland.ac.nz/external/finalproceeding/mainmenu.pdf (Accessed 8 October 2016).

Succar, B. (2009). Building information modelling framework: A research and delivery foundation for industry stakeholders. *Automation in Construction*, Vol. 18 (3), 357–75.

Tulenheimo, R. (2015). Challenges of implementing new technologies in the world of BIM: Case study from construction engineering industry in Finland. *Procedia – Economics and Finance*, Vol. 21, 469–77.

Vogt, B.A., 2010. Relating Building Information Modelling and architectural curricula. Doctoral Thesis. Kansas State University, Kansas.

Wong, A. K. D., Wong, F. K. and Nadeem, A. (2009). Comparative roles of major stakeholders for the implementation of BIM in various countries. In: Wamelink, H., Prins, M. and Geraedts, R. (eds.) *Proceedings of the International Conference on Changing Roles: New Roles, New Challenges, Noordwijk Aan Zee, The Netherlands*. Delft: TU Delft, Faculty of Architecture, Real Estate and Housing, 5–9.

Yan, H. and Damian, P. (2008). Benefits and barriers of building information modelling. In: Ren, A., Ma, Z. and Lu, X. (Eds) *Proceedings of the 12th International Conference on Computing in Civil and Building Engineering (ICCCBE XII) & 2008 International Conference on Information Technology in Construction (INCITE 2008)*, 16–18 October, 2008, Beijing.

12 Innovation roles for clients
Implementing building information modelling

Kristian Widén

Introduction

The role of information and communication technology (ICT) in the built environment, which currently focuses on building information modelling (BIM), has been a topic in construction research and the construction sector for decades. *'A "BIM Model" is a digital representation of an actual building for project communication over the whole building-project lifecycle'* (Cerovsek, 2011: 226). It means that it contains, or could contain, all necessary information to design, produce and maintain a building. Since BIM has typically been implemented one application at a time (Jung and Joo, 2011), interoperability between multiple models and multiple tools remains a key issue (Cerovsek, 2011). Furthermore, research on implementation and adoption has often focused on technical issues. This is not surprising as in ICT development, in general, the focus for a long time has been on the technical issues, especially in sectors with a strong technical orientation such as architectural, engineering and construction (AEC) industry, thus the diffusion perspective has been neglected (Peansupap and Walker, 2006a). Much of the diffused ICT innovations are of a technology push type. The developers try to market their new products (innovations) to the construction sector, which requires the adopter to fit the organisation to the technology. The success of the technology-push strategy will depend on two criteria: user accessibility and user understanding (Drury and Farhoomand, 1999). However, in the case of BIM this has not been enough to ensure industry-wide diffusion (Wamelink and Heintz, 2015).

It has been found that the success of adoption has to do with people and process as much as technical issues (Arayici *et al.*, 2011a), suggesting that there are some parts of the innovation 'picture' that are missing in the case of BIM. People need to be engaged in adoption, and their skills and understanding need to increase in order to diminish any potential resistance (Arayici *et al.*, 2011b). On a general level it is found that the complex nature of the construction process may also be a barrier to diffusion of ICT innovations (Peansupap and Walker, 2006a), owing to the particular context of construction. However, activities and development of ICT applications leading to the adoption of standards across the sector are beneficial (Miozzo *et al.*, 1998), resulting in a better and faster uptake. Similarly, implementation and diffusion can be considerably enhanced if BIM schemas and workflows are published and accessible both internally and externally (Cerovsek, 2011; Hooper, 2015).

The use and integration of ICT in the AEC industry together with the volume and value of data is expected to rise (Hooper, 2015). BIM plays an increasingly important role in shaping the future of both the construction industry and its delivered products. There has been research on the development of BIM (Pazlar and Turk, 2008) and defining its benefits (Bryde *et al.*, 2013), and some on its adoption (Linderoth, 2010). Measuring the impact of BIM has proven to be difficult as the benefits are spread among various actors in different project stages (Vestergaard *et al.*, 2011). Factors affecting BIM adoption can be divided loosely into two main areas: technical and non-technical strategic issues. The challenges lie in creating an environment which integrates both the technical and nontechnical challenges (Gu and London, 2010).

Several factors have been found to be major barriers to adoption of BIM (Gu and London 2010):

- lack of awareness and training;
- fragmented nature of the AEC industry;
- reluctance to change existing work practices;
- hesitation to learn new concepts and technologies; and
- lack of clarity on roles, responsibilities and distribution of benefits.

Little attention has been given to the role of the client in BIM adoption, and when clients have been mentioned in research it is often in terms of what others need to do in their relation to clients. For example, some argue that it is necessary for others to convince clients of the potential benefits from the use of BIM (Linderoth, 2010). On the other hand, clients need to ensure benefits on their own behalf to be able to decide and resource the implementation of BIM in their projects (Gu and London, 2010). An example of a benefit from the implementation of BIM is dynamic cost analysis (Aranda-Mena *et al.*, 2009). Other research show that clients actually have the most to gain on a general adoption of BIM (Eadie *et al.*, 2013) as well as the power to demand it (Olofsson *et al.*, 2008).

It is important to consider to what extent clients can act independently and to what extent they are influenced by their environment (Haugbølle and Boyd, 2013) as well as how willing they are to engage in or support innovation (Haugbølle *et al.*, 2015). Future research will need to address the potential for clients to influence their context as well as how to respond to it. The clients are believed to have the opportunity to impact on the rest of the industry through their procurement methods, goal setting, acting as lead user, etc., emphasising the role of the client as a change agent (Haugbølle and Boyd, 2013). Clients are believed to have a dilemma in whether to address only their own short-term goals or to take on a greater societal responsibility (Haugbølle and Boyd, 2013). In the future it will be necessary to address the issue of who will benefit from an innovation as well as how to address the risk it involves taking it beyond the traditions and norms. This chapter presents a study where a client successfully took the lead in getting BIM implemented in the construction sector on a national level. The chapter first presents some key issues on innovation, innovation implementation and diffusion in the construction sector. The next section presents the case, and this is followed by a section where the approach taken is discussed in relation to the theoretical presentation. Finally some key findings are presented.

Understanding innovation in the context of construction

Construction is seen as lagging behind many other industrial sectors. There is, however, evidence that the construction industry does address difficult and complex problems, but in ways different from other industries (Winch, 2008). In contrast to manufacturing, where new ideas may flow fairly easy through a well-established vertically integrated value-chain, construction is characterised by the way new ideas are implemented in projects where the value-chain may vary from project to project (Loosemore, 2014). The innovation process in construction cannot be modelled according to the models developed for manufacturing business sectors, due to the project-based nature of construction (Widén, 2006). For example, undertaking construction innovation outside of actual construction projects appears to be a very unusual process (Tatum, 1987) which, to some degree, often results in the development and implementation becoming parallel activities.

Understanding the key differences in regard to innovation between the construction sector and other sectors is important for understanding how existing models of innovation can be adapted or used in the context of construction (Slaughter, 1998). The organisational structure of construction with its focus on projects and temporary project organisations is an important contextual difference compared to many other sectors. If the separation between design, fabrication, implementation and many different parties are added, the contextual setting is completely different in the construction sector. Innovation in construction involves many actors within a product system and it needs to be seen as such (Blayse and Manley, 2004). This means that it is necessary to look at the complete 'picture' of construction including not only consultants and contractors, but also clients, manufacturers, regulators, etc.

Innovations often affect more than one organisation in the process making it harder for a single company to adopt something new on their own account (Miozzo and Dewick, 2004). As a result it becomes much more complex as different stakeholders need to be involved (Manseau, 2005b). As soon as an innovation affects other parties in the construction process there is need for organisational authority to ensure collaboration and integration (Slaughter, 1998). Diffusing an innovation may be frustrated if one or more affected stakeholders do not want the innovation to be adopted. During the development phase, it will be necessary to assess the likelihood of this occurrence and then to limit that possibility (Widén and Hansson, 2007). The development of a collective understanding of the innovation and creating trust at the operational level is important as it is where individuals are more likely to encounter it. A critical success factor is involvement from the early stages of development of those who will be responsible for implementation, possibly requiring mediation between new development and existing routines and duties within the organisations affected (Barlow *et al.*, 2006).

The role of clients in construction innovation

Clients are, as stated earlier, an important stakeholder for innovation in the construction sector. An important driver in itself is client dissatisfaction about current arrangements, which should be motivating enough for the construction sector to improve (Winch, 2003). Although it is clear that clients play an important role mainly in implementation and diffusion of innovation in the construction sector, few clients actually develop innovations for the rest of the sector to implement themselves. It can only be those clients who know what innovations will benefit them and their customers who can get involved, and these clients need to take an overview over the whole project and process (Brandon, 2008). Clients are also important as they shape the industry through their direct contact with the rest of the industry (Haugbølle and Boyd, 2013). It is very much the client that sets the game plan as they define the process, procurement form and requirements, which are the factors defining the boundary for other actors to relate to (Widén *et al.*, 2008). In doing so, the other stakeholders in the construction process adjust to the boundaries set by the clients. Therefore, clients are a key stakeholder of the construction process and influential on the market conditions for construction, both in relation to their own projects and as policy drivers shaping the context for the rest of the industry. Major government clients have in many cases the responsibility to use this to good effect (Barrett, 2008), formally or informally. From a theoretical perspective, Egbu (2008: 71) has defined key roles for clients to play in construction innovation (what the roles mean and what actions are related to them are elaborated in the discussion of the case):

- source/provider of knowledge;
- effective leadership;
- change agent;
- provision of financial incentives;
- appropriate forms of procurement;
- improved risk management; and
- disseminating innovations.

It is, however, important to note that clients cannot be defined as one homogenous group. There is considerable diversity between different clients (Holson and Treadaway, 2008), from non-professional one-off clients to professional repetitive clients. Not all clients have the competence and ability to drive innovation; quite few have in fact. Even if the capabilities and abilities exist, one can argue that it is not the clients' responsibility to take. Many clients are more concerned with innovations that are related to their core business rather than construction innovation (Boyd and Chinyio, 2006), which is quite natural as their core business is their livelihood. Even when they do have an interest, the risk is that engaging in innovation may prove to be too large a threat to their core business as well as to their short- and long-term financial concerns. There are also arguments that the supply side of construction has for too long not accepted responsibility for its actions, but has rather suggested that government and clients are responsible for how it acts (Holson and Treadaway, 2008).

Methods

The results are based on data gathered though interviews with people who have key insights in the issue being studied and from document studies. Interviews have been carried out both internally in the client organisation and at a strategic level with key actors from ministries and the AEC sector. The interviewees from the AEC sector were chosen from the professional and trade organisations as they were seen as having a collective view of the profession or trade they represent and from organisations and projects where the client organisation has been involved. Apart from these interviews, interviews and discussions with people at the R&D department have enriched the understanding.

To complement the interviews, the client provided internal documents, strategy plans, project reports, etc., that were studied in detail. To deepen the understanding, various other documents have been read, for example ministerial propositions and owner directives to get an understanding of the framework that the client organisation has to adhere to.

The case will be described and analysed using the key roles that were theoretically defined by Egbu (2008). The reason is threefold:

- there is a need to verify that the roles defined are roles that are important for clients to take on in relation to innovation and innovation diffusion in construction;
- it is important to gain an understanding of what actions are coupled with the different roles; and
- it gives a structure to organise the description and analysis of the case.

A client taking the lead for BIM

This section presents briefly the role the client organisation played for innovation in construction in its country. The following section presents the path the client organisation took to introduce BIM in the national construction industry and the result it generated.

The role of the client organisation for innovation in construction nationally

The national construction sector consists predominantly of small companies with a few larger companies, most of them with international owners. An earlier study has shown that the sector as a result is characterised by segmented competences and that it is necessary to connect the different knowledge areas in the sector because many innovations and implementation of innovations span many organisations and different phases in the construction process. Many of the actors are engaged in professional or trade specific organisations, for example the contractors' association, the design consultants' association or the architects' association, but also topic-specific organisations such as an organisation for those engaged in the development and use of BIM. These networks contribute to innovation through collecting the shared views of their members. Evidence from other countries shows that these types of organisation can play an important role for innovation, as pointed out by for example Gann (2000) and Winch (1998). These organisations can play the role of innovation brokers aiding in the negotiation of needed legislative changes or as knowledge contributors being one voice for a profession or trade.

The client organisation is a facilities provider and manager mainly for public clients and owned by the national state. The client organisation's main responsibility is to deliver built facilities according to user specifications and to manage the existing facilities for which they are responsible. The result is mainly measured according to cost, time and quality, where quality involves user satisfaction. The client organisation aims to be the best client in the country. The overall aims of the client organisation are built on the owner directives given by the government. Until 2008 the directives were very clear and specific on the client organisation's mission for R&D. The client organisations were to work on making the construction sector of the country more efficient, thus the ministry emphasised the importance for the client organisation to take part in R&D activities in the AEC sector. But in the directives for 2009 and 2010 these requirements were lowered, so that the client organisation was required only to report on what they have done within the field of R&D in their annual report. The client organisation, though, still considers that it has the same role as before.

Being a large client means that its actions will have an effect on innovation and development in the AEC sector at large. The effects from innovations will generally affect other organisations rather than just the innovation organisation. It is therefore important for public organisations to support the development of innovations. That in itself is one important reason for a public client such as the client organisation to shoulder the responsibility for driving the national construction sector innovation system.

In the R&D strategy for 2008–12 the client organisation was directed to contribute to the development of the construction sector through:

- using its size as a client for setting standards;

- using the client's knowledge transfer ability acquired from both building and managing the built facilities;
- using its knowledge on public demands and change processes;
- developing cooperation between actors in the sector;
- using knowledge transfer about production processes from one sector to another; and
- using its relations with national and international R&D environments.

The funds needed to carry out the tasks were taken as a percentage of the core business projects carried out. In other words, the client organisation had no public funding, but it still had a mandate (and a demand) to be in the front of the national construction sector. The client organisation also competed for funding from funding agencies the same as any other organisation nationally.

The role as informed client was mainly implemented through developing its own processes and capacity with the aim of being able to define requirements. These requirements enabled the rest of the national construction sector to develop on their own account, an approach which also allowed the client organisation to focus on the areas most valuable for it and to ensure that the development was close to its own business. There were several examples where this was used, to a large extent successfully, for example universal design, BIM and environmental solutions. One important aspect was that there were a number of other public clients, private clients and other private companies as well that followed in the footsteps of this client organisation implementing the same requirements.

Role no 1: Source/provider of knowledge

The client organisation was clearly a source as well as a provider of knowledge. When the client organisation did not have the knowledge necessary it made sure it found the relevant knowledge. Early on, even before the project of implementing BIM was started, the client organisation engaged in both national and international organisations addressing BIM development. The client organisation wanted to gain as much knowledge as possible, but it was also interested in influencing the path the development took, ensuring that it had something to gain from the development of BIM. In these organisations, both internationally and nationally, they were open with and shared their understanding and knowledge.

The client organisation developed its knowledge in the early implementation projects. The lessons learned and the routines they developed were disseminated both within their own organisation and others interested in the rest of the industry. Through the chosen approach it also managed to address the factors Gu and London (2010) found to be major barriers to adoption of BIM. The client organisation trained people both from their own organisation as well as from other organisations. It also made the industry aware of BIM and its development and implementation.

Role no 2: Effective leadership

According to Kotter (1996), effective leadership in relation to innovation consists of three components:

- establishing direction;
- aligning people; and
- motivating and inspiring.

The client organisation showed leadership in all three. Together with other national public organisations, it very early on got involved in the development of BIM internationally as described above. In this work it decided to pursue and promote the use of an open standard BIM – using IFC – thus clearly showing direction. The main reason was the risk of a lock-in effect and otherwise becoming dependent on a sole supplier. In 2005 the client organisation announced that from 1 January 2010 BIM should be implemented in all projects. This caused uproar in the national construction industry. Very few thought that this would be possible. IFC was not developed enough at that time, there was a lack of tools to be used, etc. The knowledge of BIM, how to use it and consequences of its use was lacking in the industry. There was a general unease from the construction industry, as well as some internal parts of the client organisation. Still, from 1 January 2010 BIM has been used successfully, although not entirely without problems. Very soon after, some larger public organisations also adopted the guidelines. Today the guidelines have developed further, and more and more client organisations, both public and private, are using them. The reason behind the success was the well-planned and thought-out process on how to move from the 2005 situation in order to reach the goal in 2010.

The early choice made by the client organisation of an open standard shaped much of the process. First of all, the choice was made on the assumption that it would be the best choice for the client itself, from both short- and long-term perspectives. The industry as a whole did not think it was the right choice, but since the client organisation stuck with it the rest of the industry had to comply. The development and implementation could from its outset focus on one system, or rather any system that could communicate in IFC. This meant that everyone knew the 'rules of the game' early on, whether they liked it or not. The necessary technical support was developed. But the main point was that it was possible from an early stage to leave the technical focus and focus on developing standards which have been identified as very important for a better and faster uptake both for ICT in general (see e.g. Miozzo *et al.*, 1998) and for BIM in particular (see e.g. Hooper, 2015; Cerovsek, 2011). This was clearly a motivational factor for the rest of the industry. The chosen process also meant a step away from the traditional technology push towards a demand pull type of innovation as the client organisation as well as other actors had asked for specific solutions. The process also meant a step away from the technical focus and a much stronger focus on the people that were going to use BIM.

There is also strong evidence of successful interaction with the construction sector during the development, aligning people and organisations, thus leading to the new requirements that would allow parties working with the client organisation to learn how to meet the new requirements. This has been done on the sector level involving many parties, for example professional organisations, through cooperation in the organisation for those interesting in the development and use of BIM, and through development projects where specific actors have been able to develop, for example, tools that the rest of the industry can make use of.

Role no 3: Change agent

The client organisation was definitely a change agent in that it included the entire construction sector and persuaded them to change. Egbu (2008) notes that being a change agent involves, for example, influencing industry structures, culture and reward systems, influencing the market to implement an innovation, etc. The client organisation has a strong position in its national context and it also has an important role in driving innovation in the national construction sector. As such, the client organisation has the potential to affect the rest of the industry much in line with what has been found possible for clients to do by, for example, Haugbølle and Boyd (2013), Barrett (2008) and Widén *et al.* (2008). In this particular case, the client organisation did it completely from the perspective of what would provide the greatest benefits to the organisation itself and its customers. Linking BIM to facilities management was left for future development. However, the client did so consciously with the aim of limiting the complexity and from the perspective that, compared to the pre-BIM process, the information would at least be as good as before BIM was implemented.

By focusing on the design and production phases of the construction process the client organisation could ensure that all types of parties being affected were involved as addressed by Blayse and Manley (2004). The client organisation did so in a number of ways including interactions with national professional, trade and topic-specific organisations, with companies through education schemes, with specific companies that took part in the development and implementation in the early projects, and with others through joint ventures developing the specific innovations. With this approach the client organisation managed to address the very complex stakeholder situation found troublesome in much of the earlier research by, for example, Manseau (2005b), Miozzo and Dewick (2004) and Slaughter (1998).

The client organisation made people and organisations change their existing work practices. The companies in the construction sector had to do this if they wanted to keep doing business with the client organisation, which also meant that they had to learn new concepts and technologies, but the client organisation provided the tools, and to some extent the means, for them to do it. The roles and responsibilities were made clear, as was the distribution of benefits.

Role no 4: Provision of financial incentives

Incentives are most often seen as a means to award a particular solution, approach or path taken. The lack of incentives to innovate is generally seen as one important obstacle to innovation in construction (Blayse and Manley, 2004). The client organisation provided a financial incentive in the sense that the client organisation said that all projects after 1 January 2010 would be through BIM. Any organisation wanting to do business with them needed to be prepared. The financial incentive, however, would not give a premium to suppliers implementing BIM; rather there would be no business for them unless they had implemented BIM.

Apart from that, development of tools and other research was financed directly and indirectly by the client organisation. The focus of these projects was either on projects that were of strategic importance, for example the development of a BIM manual, or the development of tools necessary to succeed with the implementation. In the early implementation projects the client organisation did fund the additional cost of the projects, so as to not put any unnecessary burden on their suppliers. But there was the perspective that in-house costs to develop the capabilities among its suppliers had to be covered by the suppliers themselves as part of their own development.

Role no 5: Appropriate forms of procurement

The client has a good opportunity to influence its suppliers through the choice of procurement approach. Public procurement especially has the potential to assert incentives in a favourable direction (Albano *et al.*, 2013). Much focus in construction research has been on using procurement forms that allow, as well as promote, development and implementation of innovation (see e.g. Courtney, 2008). In so doing the client is seen to leave the innovation choice as well as risk to the market. Consequently, an innovation such as BIM could have been encouraged through the use of procurement methods with incentives to implement BIM rather than taking charge of the whole process. The client organisation chose not do this for two reasons:

- They wanted to ensure that they got the full benefit of using BIM in its projects. This included the promotion of openBIM/IFC as well as a BIM manual that supported its business processes.
- They had taken a strategic decision a few years earlier to have a strict project process with stage gates and a clearly formulated procurement process which they did not want to change. The result was that both the project management process and the procurement process stayed the same, but with BIM as the information carrier instead of traditional design drawings.

As a result, there was not much done to change the forms of procurements as such, but all standard documents used were adapted to the use of BIM. In addition to this the developed and implemented BIM manual was to be used in all projects. The client organisation, then, did not use the procurement approach to 'make' the construction sector develop BIM on its behalf, but rather made sure that the procurement approach would not pose any obstacle to the implementation.

Role no 6: Improved risk management

The client organisation managed risk in a number of ways, for itself as well as for the industry as a whole. Introducing an innovation is always linked to risks especially if the innovation is not fully developed, as was the case here. The extra risks as well as the development work meant increased cost that needed to be covered. The approach the client organisation took in developing and implementing its BIM approach was to first implement it in one project. With the result, and from lessons learned in the first project, the approach was implemented in 5 projects for further testing. It was then validated in 15 projects and finally implemented generally in all projects. In the first two stages the responsibilities for implementation and also for the extra costs were allocated to the development department. In the second to last and final step the responsibility was transferred to the construction supervision department of the client organisation (i.e. the department normally responsible for carrying out projects). In the general implementation stages the development department had a supporting responsibility that after some time was transferred to the professional services department.

As the client organisation focused on an inclusive process, engaging organisations from all parts of the construction process, it also decreased the risk of having stakeholders hindering the implementation (see e.g. Widén and Hansson, 2007). The client organisation instead managed to involve the actors and actual people responsible for the implementation at various levels and organisations, which is very important as argued by Barlow *et al.* (2006). By doing this, the client organisation managed to build a collective understanding of BIM as well as building trust for it at the operational level, which was necessary to achieve the success the client organisation did. One other thing the client organisation did was to focus on the delivery and design phase as described earlier. By limiting the complexity the risks were also limited.

Role no 7: Disseminating innovations

The client organisation disseminated the innovations it had developed as well as those developed by others on their behalf in several ways: as part of the implementation projects, through reports and seminars with stakeholders, etc., teaming up with private partners in joint ventures in order to develop specific innovations. Examples include software development and 'demand BIM' (a tool allowing project demands to be introduced in the BIM model). Some of these developments are on the verge of becoming national as well as international standards. Another important positive factor was that the client organisation made its research and development available for the rest of the sector to learn from as well as routines and manuals.

Almost immediately after the full implementation from 1 January 2010 a number of other public clients at various levels adopted the BIM manual with success. To this date that number has increased and a number of private clients have now also adopted it. The BIM manual has been revised and there is ongoing development on how to use the information generated in the project-specific BIM for the facilities management.

Conclusion

The approach the client organisation took for implementing BIM meant it had a number of advantages compared to simply letting the rest of the industry take charge. As a change agent, the client organisation chose the BIM system, namely IFC open BIM, which best suited them. By choosing this approach the client organisation ensured that it would gain from driving this change.

The client organisation focused on implementing BIM for the delivery of projects so that the scope would not be too large. Some of the key activities of the client organisation from announcing the intent to implement IFC open BIM and the actual implementation included:

- workshops initially with the stakeholder from industry to gain an understanding of the different perspectives that the stakeholders had on the implementation;
- financing the development of tools to ensure that the lack thereof would not hamper the implementation;
- arranging teaching and training programmes, both internally and externally, so that all parties would be prepared;
- running a number of test implementations with the aim of testing and learning, the '1–5–15–all' projects approach; and
- parallel to the other activities developing the BIM manual.

By announcing well in advance their intention to demand BIM gave both the client organisation itself and the rest of the industry the time to adapt. It also allowed the client organisation to carry out the key activities described before the time of full implementation. This way it could eliminate a number of different risks. The client organisation ensured through continuous dialogue with the different actors of the industry that the industry would be as prepared as possible. This way, almost all types of risk normally associated with the implementation of such a complex innovation were kept to a minimum. This work was of course not without cost. The majority of the cost was financed in-house, but the client also ensured some external finance. Due to the lower risk of implementation and the prospect of getting an approach that would better suit the client, it was considered a necessary cost.

Following Egbu (2008), this research shows that the seven innovation roles of clients are valid, at least in this case, and with the client organisation taking more or less all of those roles during the development and implementation of BIM. Because of the specific context of the client organisation, the nature of the innovation and the approach taken by the client organisation, some roles are more prominent than others. The roles that were most visible are those of being a strong leader and a change agent, and these two are interlinked. Having a privileged position in the national construction system, it may not be surprising that the client organisation clearly took on these two roles. A general lesson is that the other roles and how they were executed were very much a result of the client organisation shouldering the roles of being strong leader and change agent and the decisions those lead to.

Disseminating knowledge, risk management and innovations were also important and clearly visible. Giving financial incentives and using appropriate procurement methods were not as prominent. This does not suggest that they were not important, though. If they had not been dealt with appropriately to the degree necessary the result would probably not have been successful. Being dealt with appropriately meant not necessarily using the prevailing understanding of how to do it, but rather in a way that best suited the particular innovation and the context of the client organisation. This again falls back to the client's role as a strong leader and change agent with the decisions that followed from this.

One important lesson is that nothing of this would have happened if the client had not seen the benefit of implementing BIM. All actions were taken with attention to what would be most beneficial for the client organisation now and for the future. Another important lesson from this study is that, although it is beneficial to use this framework to understand how clients may influence innovation in construction, it is through the actions the client took in relation to these roles that actually decided whether they are successful or not. It had to be done through the particular understanding of the innovation and the context in which the client was operating. Future research should investigate in what ways the actions of clients in other contexts align with these seven innovation roles and support innovation and innovation implementation.

References

Albano, G. L., Snider, K. F. and Thai, K. V. (2013). Charting a course in public procurement innovation and knowledge sharing. In: Albano, G. L., Snider, K. F. and Thai, K. V. (Eds.). *Charting a course in public procurement innovation and knowledge sharing*. Boca Raton, FL: PrAcademics Press, 1–27.

Aranda-Mena, G., Crawford, J., Chevez, A. and Froese, T. (2009). Building information modelling demystified: Does it make business sense to adopt BIM? *International Journal of Managing Project in Business*, Vol. 2 (3), 419–34.

Arayici, Y., Coates, P., Koskela, L. Kagioglou, M., Usher, C. and O'Reilly, K. (2011a). Technology adoption in the BIM implementation for lean architectural practice. *Automation in Construction*, Vol. 20 (2), 189–95.

Arayici, Y., Coates, P., Koskela, L. Kagioglou, M., Usher, C. and O'Reilly, K. (2011b). BIM adoption and implementation for architectural practices. *Structural Survey*, Vol. 29 (1), 7–25.

Barlow, J., Bayer, S. and Curry, R. (2006). Implementing complex innovations in fluid multi-stakeholder environments: Experiences of 'telecare'. *Technovation*, Vol. 26 (3), 396–406.

Barrett, P. (2008). A global agenda for revaluing construction: The client's role. In: Brandon, P. and Lu, S.-L. (Eds.). *Clients driving innovations*. Chichester: Wiley-Blackwell, 3–15.

Blayse, A. M. and Manley, K. (2004). Key influences on construction innovation. *Construction innovation*, Vol. 4 (X), 143–54.

Boyd, D. and Chinyio, E. (2006). *Understanding the construction client*. Oxford: Wiley-Blackwell.

Brandon, P. (2008). Preface. In Brandon, P. and Lu, S.-L. (Eds.). *Clients driving innovation*. Chichester: Wiley-Blackwell, xv–xxii.

Bryde, D., Broquetas, M. and Volm, J. M. (2013). The project benefits of Building Information Modelling (BIM). *International Journal of Project Management*, Vol. 31 (7), 971–80.

Cerovsek, T. (2011). A review and outlook for a 'Building Information Model' (BIM): A multi-standpoint framework for technical development, *Advanced Engineering Informatics*, Vol. 25 (2), 224–44.

Courtney, R. (2008). Enabling clients to be professional. In: Brandon, P. and Lu, S.-L. (Eds.). *Clients driving innovation*. Chichester: Wiley-Blackwell, 33–42.

Drury, D. H. and Farhoomand, A. (1999). Information technology push/pull reactions. *Journal of Systems and Software*, Vol. 47 (1), 3–10.

Eadie, R., Browne, M., Odeyinka, H., McKeown, C. and McNigg, s. (2013). BIM implementation throughout the UK construction project lifecycle: An analysis. *Automation in Construction*, Vol. 36, 145–51.

Egbu, C. (2008). Clients' roles and contributions to innovations in the construction industry: When giants learn to dance. In: Brandon, P. and Lu, S.-L. (Eds.). *Clients driving innovation*. Chichester: Wiley-Blackwell, 69–77.

Gann, D. (2000). *Building innovation: Complex constructs in a changing world*. London: Thomas Telford Publishing.

Gu, N. and London, K. (2010). Understanding and facilitating BIM adoption in the AEC industry. *Automation in Construction*, Vol. 19 (X), 988–99.

Haugbølle, K. and Boyd, D. (2013). *Clients and users in construction: Research roadmap report*. CIB Publication 371. Rotterdam: CIB.

Haugbølle, K., Forman, M. and Bougrain, F. (2015). Clients shaping construction innovation. In: Orstavik, F., Dainty, A. and Abbott, C. (Eds.). *Construction innovation*. Chichester: Wiley Blackwell, 119–33.

Holson, J. and Treadaway K. (2008). Is the client really part of the team? A contemporary policy perspective on Latham/Egan. In: Brandon, P. and Lu, S.-L. (Eds.). *Clients driving innovation*. Chichester: Wiley-Blackwell, 26–32.

Hooper, M. (2015). *BIM anatomy II: Standardisation needs and support systems*. Lund: Lund University.

Jung, Y. and Joo, M. (2011). Building information modelling (BIM) framework for practical implementation. *Automation in Construction*, Vol. 20, 126–33.

Kotter, J. P. (1996). *Leading change*. Boston: Harvard Business School Press.

Linderoth, C. J. (2010). Understanding adoption and use of BIM as the creation of actor networks. *Automation in construction*, Vol. 19 (1), 66–72.

Loosemore, M. (2014). *Innovation strategy and risk in construction: Turning serendipity into capability*. London: Routledge.

Manseau, A. (2005b). Redefining innovation. In: Manseau, A. and Shields, R. (Eds.). *Building tomorrow: Innovation in construction and engineering*. Aldershot: Ashgate, 43–55.

Miozzo, M. and Dewick, P. (2004). *Innovation in construction*. Cheltenham: Edward Elgar.

Miozzo, M., Betts, M., Clark, A. and Grilo, A. (1998). Deriving an IT-enabled process strategy for construction. *Computers in Industry*, Vol. 35 (1), 59–75.

Olofsson, T., Lee, G. and Eastman, C. (2008). Editorial: Case studies of BIM in use. *ITcon*, Vol. 13, 244–5.

Pazlar, T. and Turk, Z. (2008). Interoperability in Practice: Geometric data exchange using the IFC standard. *ITcon*, Vol. 13, 362–80.

Peansupap and Walker (2006a). Information communication technology (ICT) implementation constraints: A construction industry perspective. *Engineering, Construction and Architectural Management*, Vol. 13 (4), 364–79.

Slaughter, S. (1998). Models of construction innovation. *Journal of Construction Engineering and Management*, Vol. 124 (3), 226–31.

Tatum, C. B. (1987). Process of innovation in construction firms. *Journal of Construction Engineering and Management*, Vol. 113 (4), 648–63.

Vestergaard, F., Karlshøj, J., Hauch, P., Lambrecht, J. and Mouritsen, J. (2011). *Måling af økonomiske gevinster ved Det Digitale Byggeri. Rapport SR 12–02—SR 12–07 (in Danish: Measuring economic benefits of the development programme Digital Construction)* Lyngby: DTU BYG, Technical University of Denmark.

Wamelink, J. W. F. and Heintz, J. L. (2015). Innovating for integration. In: Orstavik, F., Dainty, A. and Abbott, C. (Eds.). *Construction Innovation*. Chichester: Wiley Blackwell, 149–64.

Widén, K. (2006). *Innovation Diffusion in the Construction Sector*. Lund: Division of Construction Management, Lund University.

Widén, K. and Hansson, B. (2007). Diffusion characteristics of sector financed innovation. *Construction Management and Economics*, Vol. 25 (5), 467–76.

Widén, K., Atkin, B. and Hommen, L. (2008). Setting the game plan: the role of clients in construction innovation and diffusion. In: Brandon, P. and Lu, S.-L. (Eds.). *Clients driving innovation*. Chichester: Wiley-Blackwell, 78–87.

Winch, G. (1998). Zephyrs of creative destruction: Understanding the management of innovation in construction. *Building Research and Information*, Vol. 26 (5), 268–79.

Winch, G. (2003). How innovative is construction? Comparing aggregated data on construction innovation and other sectors – a case of apples and pears." *Construction Management and Economics*, Vol. 21(6), 651–4.

Winch, G. M. (2008). Revaluing construction: Implications for the construction process. In: Brandon, P. and Lu, S.-L. (Eds.). *Clients Driving Innovation*. Chichester: Wiley-Blackwell, 16–25.

13 Client innovation networks

Kim Haugbølle, Stefan Christoffer Gottlieb,
Niels Haldor Bertelsen and Peter Vogelius

Introduction

The significance of public procurement is increasingly being recognised as a potential source of innovation and one of the key elements of a demand-oriented innovation policy as pointed out by among others Edler and Georghiou (2007) and Hommen and Rolfstam (2009). Hence, in recent years pre-commercial procurement (PCP) and innovation partnerships have become part of innovation policy strategies of the European Union and several member states. Innovation partnerships are a public procurement procedure established in the latest version of the EU services directive (European Union, 2014), while pre-commercial procurement refers only to procurement of R&D services and is considered an exemption that falls outside the recent services directive.

A number of studies on the role of construction clients in innovation have been published in the past decade, notably through the conference series 'Client's driving construction innovation' (see Brown *et al.* 2005, 2006, 2008) and the subsequent book edited by Brandon and Lu (2008) addressing the context for innovation, the innovation process and how to move ideas into practice. In particular public clients are frequently being encouraged to take the lead in a transition of the construction industry towards providing better value for money, striving towards a sustainable built environment, improving productivity, etc. However, moving from policy calls to practical action requires a better understanding of the key issue of how construction clients may take on this lead as pointed out by the CIB W118 roadmap on clients and users in construction (Haugbølle and Boyd, 2013, 2016).

One example is Miller (2008), who points at large engineering projects as joint innovations between clients and consultants. Another example includes Haugbølle *et al.* (2015), who develop a typology of clients based on their inclusion in different technological frames of either production or consumption. While these studies focus mainly on the impact of clients' activities on developing new practices in the construction industry, few studies take a look at the inner operation of client organisations and how they deal with innovation and change internally. Notable exceptions include Hoezen (2012) on the Dutch Highways Agency's adoption and adaptation of competitive dialogue as procurement method, the 36 case studies of best practices and changes in facility management organisations (Jensen *et al.*, 2008) and a change management study of a Danish urban renewal organisation (Haugbølle and Gottlieb, 2007). Most studies of clients tend to treat clients as singular entities, while very few studies look at clients in plurality. One very notable exception is Courtney (2008), who took the lead in establishing the International Construction Clients Forum (ICCF) on behalf of CIB – the International Council on Research and Development in Building and Construction. Although individual clients may stimulate change in the construction industry, sustained and systemic change is more likely to take place if clients join forces to alter those mechanisms and practices that stifle innovation.

This chapter analyses how clients can organise themselves in innovation networks, establish collaboration with private and public funding bodies, and manage a range of development activities. This chapter assesses the impacts and benefits of clients forming such innovation networks and it highlights some critical dilemmas and challenges that client innovation networks may face. It will do so by re-analysing and discussing lessons learned from previous evaluations of three different client innovation networks in Denmark and Sweden carried out by the authors of this chapter.

Methodology

Transition theory

The question of how client innovation networks may stimulate and manage sociotechnical dynamics in construction requires a theoretical perspective that is capable of addressing both change and stability at micro, meso and macro levels simultaneously in the short as well as the long term. Hence, this chapter adopted a multilevel perspective to re-analyse the three innovation networks. This multilevel perspective combines insights from constructivist science and technology studies with evolutionary economics (see e.g. Rip, 1995; Rip and Kemp, 1998; Kemp *et al.*, 1998; Kemp *et al.*, 2001; Geels, 2002). The multilevel perspective applies the concepts of niches, regimes and landscapes in order to understand the dynamics of change and stability (Geels, 2002):

- *Niches* form the micro-level where novel technologies emerge and develop or disappear, usually in a rapid pace.
- Sociotechnical *regimes* form the meso-level, which accounts for the stability of existing large-scale systems.
- Sociotechnical *landscapes* form the macro-level (e.g. macro-economics, deep cultural patterns, macro-political developments) beyond the direct influence of niche and regime actors. Macro-level changes usually take place slowly, in the order of decades.

The concept of sociotechnical regimes constitutes the centrepiece of the multilevel perspective. This concept builds on the concepts of technological

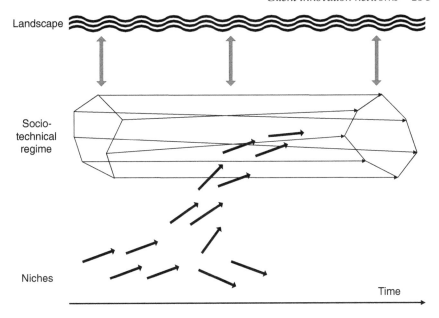

Landscape

Socio-
technical
regime

Niches

Time

Figure 13.1 A dynamic multilevel perspective on transitions

Adapted after Geels, 2002: 1263

regime (Nelson and Winter, 1982) and technological trajectories (Dosi, 1982) from evolutionary economics, but extends the inherent focus on engineers and technologists to other groups such as policy makers, users and special-interest groups as these actors also shape sociotechnical development as has been pointed out by sociologists of technology (see e.g. Bijker, 1995 and Oudshoorn and Pinch, 2003). Geels (2002) identifies seven dimensions in a sociotechnical regime: technology, user practices and application domains (markets), symbolic meaning of technology, infrastructure, industry structure, policy and techno-scientific knowledge (Figure 13.1).

The central argument of this multilevel perspective is that changes or transitions come about through interactions between processes at three different levels:

- innovations in the niches build up internal momentum;
- changes at the landscape level create pressure on the regime; and
- destabilisation of the regime creates windows of opportunity for niche innovations.

Sociotechnical transitions can take place through at least four different pathways:

- transformation;
- technological substitution;

- reconfiguration; and
- de-alignment and re-alignment.

(Geels, 2002)

Table 13.1 provides an overview of the four different transition pathways.

In the remainder of the chapter, the transition theoretical framework will be used to explicate how the client innovation networks strategically organise their activities in order to act as change agents. In particular, emphasis is placed on illustrating how the client innovation networks are able to manage tensions between processes at micro, meso and macro levels, thus contributing to change and innovation in the construction industry.

Table 13.1 Typology of sociotechnical transition pathways

Transition pathways	Main actors	Type of (inter)actions	Keywords
1. Transformation	Regime actors and outside groups (social movements).	Outsiders voice criticism. Incumbent actors adjust regime rules (goals, guiding principles, search heuristics).	Outside pressure, institutional power struggles, negotiations, adjustment of regime rules.
2. Technological substitution	Incumbent firms versus new firms.	Newcomers develop novelties, which compete with regime technologies.	Market competition and power struggles between old and new firms.
3. Reconfiguration	Regime actors and suppliers.	Regime actors adopt component-innovations, developed by new suppliers. Competition between old and new suppliers.	Cumulative component changes, because of economic and functional reasons. Followed by new combinations, changing interpretations, and new practices.
4. De-alignment and realignment	New niche actors.	Changes in deep structures create strong pressure on regime. Incumbents lose faith and legitimacy. Followed by emergence of multiple novelties. New entrants compete for resources, attention and legitimacy. Eventually one novelty wins, leading to re-stabilisation of regime.	Erosion and collapse, multiple novelties, prolonged uncertainty and changing interpretations, new winner and re-stabilisation.

Source: Adapted after Geels and Schot, 2007: 414

Research design

This chapter is based on four previous evaluations of three different client innovation networks in Denmark and Sweden. These innovation networks include:

- the Danish network AlmenNet for the Danish social housing sector covering a broad range of development activities;
- the Swedish BeLok network for energy savings in commercial buildings; and
- the Swedish BeBo network for energy savings in the residential sector. This network was studied twice with a five-year interval.

It should be noted that each of the four studies differs with regard to theoretical grounding, research approach, methods and data. The two analyses of BeBo adopted the same business strategy perspective, while the two studies on AlmenNet and BeLok did not adopt an explicit theoretical framework. The research approach was analytical with regard to the evaluations of BeLok and BeBo, while the research approach to AlmenNet was action-oriented with one of the authors being the network manager. The studies of the three networks applied different methods for data collection and sources of data. The evaluations of BeBo and BeLok included qualitative research interviews with board members and secretariats, documentary analysis of policy and strategy documents, websites, reports from development activities and discussions with representatives of the networks and funding agency of the conclusions reached in the evaluations. The evaluation of AlmenNet had access to the same type of data, but in addition also enjoyed the benefit of one of the authors having been an observant participant throughout the entire process of raising funds for the network, establishing the network and operating the network for several years.

Case 1: The BeBo network for energy-savings in housing

The following section summarises two previous evaluations by Gottlieb and Haugbølle (2010) and Haugbølle and Vogelius (2016). Both evaluations were conducted using the same business strategy framework developed by Hambrick and Fredrickson (2001). Table 13.2 provides an overview of BeBo.

Background

BeBo (Beställargruppen Bostäder) was established in 1989 as a procurement group for housing by the Swedish Energy Agency (Energimyndigheten) in collaboration with the largest group of Swedish residential property owners. Since 2005 BeBo has operated as a network under the auspices of Byggherrarna (Swedish Construction Clients). The network has been instrumental in efforts by the Swedish government to reduce energy consumption and improve energy efficiency in the built environment.

Table 13.2 Overview of BeBo

	BeBo
Country	Sweden
Established	1989
Scope	Technology procurement
Themes	Energy savings
Funding organisation	Public agency
Co-funding	Min. 50 %
Membership	Residential property owners
Secretariat	Part-time
	Client association plus consulting firm
Activities	Demonstration projects
	Development of tools etc.
	Dissemination

BeBo (2014) defined its mission and prime objectives as follows: 'Through the development of common procurement skills, the activities of BeBo shall help accelerate earlier introduction of energy-efficient systems and products on the market.' This entailed the following objectives:

- Conduct investigations and measurements to elucidate potentials.
- Demonstrate and evaluate new solutions.
- Conduct feasibility studies as a basis for technology procurement.
- Implement technology procurement.
- Market and introduce energy-efficient technology.
- Identify and disseminate lessons learned.
- Act as a sounding board for the Swedish Energy Agency and other agencies within the network's areas of expertise.

The public funding of BeBo amounted to SEK 9 million (EUR 1.2 million) in 2009 and SEK 32 million (EUR 4.25 million) for the four-year period 2012–15. The public funding was to be geared with at least the same amount of funding from the members of the network.

Organisation

The organisation of the client innovation network was highly formalised via formal membership, a board and a secretariat. At the first evaluation in 2009, BeBo had 20 members: social housing associations, public authorities and professional organisations. Since 2009, BeBo had recruited a handful of new members. Although BeBo was widely known, the membership base tended to stay rather static. Hence it appeared as if the membership base essentially covered most of the potential relevant members.

The BeBo board was composed of a chairman, the previous secretary of the network, representatives of the Swedish Energy Agency and the Swedish National Board of Housing, Building and Planning (Boverket), along with representatives of the Swedish Construction Clients and the Swedish Association of Public Housing Companies (SABO). The secretariat was divided in two: an administrative section hosted by Swedish Construction Clients, and a technical section hosted at a large consulting engineering company. Both sections were based on part-time secretariat staff. While this division ensured a competent secretariat, it also entailed a dilemma with regard to operational flexibility versus the cross-pressure of other commitments in the home organisation of the staff members.

Activities and achievements

The activities of BeBo comprised three main types:

- demonstration projects;
- tools development; and
- dissemination.

BeBo initiated development projects especially revolving around technology procurement, which could be viewed as the primary vehicle of strategic realisation.

The prime approach of BeBo was the extensive use of demonstration projects. While some demonstration projects had a rather narrow focus on specific topics such as heat recovery of ventilated air (see e.g. Wahlström, 2014), others had a more comprehensive perspective where several energy-saving initiatives were applied and tested as 'packages'. This was particularly prominent in the approach of the programme of Rekorderlig Renovering ('Record Breaking Refurbishment') with the ambition of reducing energy consumption by 50 per cent by applying a comprehensive array of measures. The approach was tested in five parallel demonstration projects distributed regionally (Levin and Larsson, 2012).

Other development activities included the development of a tool for profitability calculations ('BeBo's lönsamhetskalkyl') and the so-called 'Godhetstal' – key performance indicators for obtaining a level of minimum performance or best practice performance. Further, BeBo worked on developing a standardised refurbishment process model with various tools, guidelines, etc., based on incorporating lessons learned through 'Record Breaking Refurbishment'.

A third main activity was dissemination of knowledge. The comprehensive campaign 'Halvere Mera' ('Halving More') aimed at disseminating the methods and approach from 'Record Breaking Refurbishment' in three steps:

- pre-studies;
- implementation; and
- closure.

The 2010 ambition was to reach 100 new renovation projects before 2015 (Högdal, 2013). The campaign succeeded in starting 31 new pre-studies and 17 smaller energy audits within a six-month period. Although the campaign did not reach 100 projects, it managed to reach a different subset of actors that otherwise tended to be difficult to get engaged. The examples and results of the campaign are summarised in Högdal (2013).

From 2009 to 2014 BeBo greatly improved its dissemination activities. Hence, BeBo prepared a new communication strategy, employed a part-time communication manager, introduced regular reporting to the board, upgraded the BeBo website and improved web statistics. The new communication approach was multisided. Although the website was regarded as the vital communication platform, it was added to by interviews, oral presentations at conferences, seminars and other meetings all over Sweden, as well as contributions, articles and letters in professional trade journals.

Findings

In line with the business strategy analysis, the strategy of BeBo was assessed along six criteria:

- The evaluations stressed that the strategy of BeBo was well-aligned with political agendas at both national and European level. Similarly, BeBo was well-aligned with the business environment due to the focus on commercial implementation and the focus on cost-effective energy savings rather than maximum energy savings. With regard to the context of use and operation, some issues related to the use phase were addressed, but user behavioural and managerial issues had a more marginal position.
- The evaluations stated that BeBo exploited its limited resources well through close collaboration with national research institutions.
- The evaluations characterised the differentiators of BeBo as being strong due to the close ties with the Swedish Energy Agency, its non-commercial purpose and its evidence-based approach to documenting the results of demonstration projects. However, the differentiators were deemed sensitive to changes in governmental financial support. Further, BeBo largely left the use of buying power and purchasing volume of BeBo members unexplored and unexploited as a differentiator.
- The match between the different elements of BeBo's strategy was considered appropriate for the execution of the strategy despite the somewhat broadly formulated mission and objectives along with some internally unmediated challenges and dilemmas. These included among others the classical dilemma of long lead times of renovation projects versus the typical shorter time scale for development programmes as well as the upscaling from individual projects to entire portfolios in client organisations.
- The evaluations pointed at two resource constraints. First, technology procurement of new integrated solutions ('packages') such as façade

insulation systems proved to be very challenging for BeBo. Second, addressing new behavioural, managerial and financial issues more firmly would either require additional resources or a shift in prioritisation of available resources.

- The evaluation questioned whether the strategy of BeBo with its ambitious targets was implementable. Although some demonstration projects pointed at the possibility of achieving a 50 per cent energy reduction, implementation through technology procurement and/or dissemination of lessons learned were not considered sufficiently effective without the use of additional measures such as stricter regulation.

Case 2: The BeLok network for energy savings in commercial facilities

The following section summarises a previous evaluation of BeLok published as an internal document (Bertelsen, 2008) and as a conference proceeding (Haugbølle and Bertelsen, 2008). Table 13.3 provides an overview of BeLok.

Background

BeLok (Beställargruppen Lokaler) was established in 2001 as a procurement group for commercial facilities by the Swedish Energy Agency (Energimyndigheten) in collaboration with the largest Swedish commercial property owners and administrators. BeLok operated as a network under the auspices of Byggherrarna (Swedish Construction Clients). As innovative purchasers the network was instrumental in the efforts by the Swedish government to reduce energy consumption by 20 per cent in 1995–2020 and 50 per cent over the whole period 1995–2050 through technology procurement and dissemination of energy efficient technologies.

Table 13.3 Overview of BeLok

	BeLok
Country	Sweden
Established	2001
Scope	Technology procurement
Themes	Energy savings
Funding organisation	Public agency
Co-funding	Min. 50 %
Membership	Commercial facilities owners
Secretariat	Part-time
	Client association plus university
Activities	Demonstration projects
	Development of tools etc.
	Dissemination

The vision, mission and objectives of BeLok were:

- Execute development and dissemination activities to reduce energy consumption and improve functionality and comfort in commercial facilities.
- Support promising and energy efficient products, systems and methods and create the necessary conditions for fast implementation and market diffusion.
- Secure cost effective energy solutions balancing long-term costs with energy reductions.
- Follow up by evaluation and monitoring the realised energy improvements and the effect on the building performance.

Since 2005, BeLok has been responsible for drafting two-year development programmes and selecting relevant development projects to be initiated. The programme covering the period 2007–8 had a total budget of EUR 2.6 million (SEK 25 million), of which EUR 0.9 million (SEK 9 million) was granted by the Swedish Energy Agency and the remaining 64 per cent was financed by BeLok members.

Organisation

The organisation of the client innovation network was highly formalised via formal membership, a board and a secretariat. In 2007, 13 commercial real estate owners were members of BeLok, of which 8 had been members since the establishment of BeLok. The membership base comprised five private companies, two companies owned by regions and six companies owned by the Swedish government. Although the number of members was fairly small, the BeLok members represented close to 20 per cent of the market for commercial facilities. Still, the evaluation suggested increasing the number of members.

The BeLok board had eight members: a chairperson (a professor from Chalmers University of Technology) and seven members including the chairman of Swedish Construction Clients and a representative of the Swedish Energy Agency. Since 2005, the network has been hosted by Swedish Construction Clients. The secretariat was divided in two: an administrative section hosted by Swedish Construction Client, and a technical section hosted by Chalmers University of Technology. Both sections were based on part-time secretariat staff. While this division ensured a competent secretariat, it also entailed a dilemma with regard to operational flexibility versus the cross-pressure of other commitments in the home organisation of the staff members.

Activities and achievements

The activities of BeLok covered development and dissemination. With regard to development projects, the majority of these were primarily based on 'a good idea'. This practice, however, had a number of implications. First, the type of innovations addressed by BeLok tended to be incremental in order to reduce the associated risks for the participants. Second, the development projects broadly spanned two categories: technical product-oriented projects, and process-related development projects. Technical projects dominated the portfolio of development projects in terms of both numbers and budget. BeLok put less effort into the process-related projects compared with the technical product-oriented projects. Third, certain perspectives such as the perspective of end-users were largely left out of sight with regard to the aim of the projects, who participated in the projects and the target groups for disseminating lessons learned.

With regard to dissemination activities, internal meetings aimed at the members of BeLok was the predominant method although some other dissemination activities were also done. Frequent interaction was assured by two-day meetings four times a year, where ideas were turned into action and development projects via informal discussions. While commitment and frequent interaction characterised the internal dissemination of results, the external dissemination to actors outside BeLok tended to be more random. An important platform was the BeLok's website, with free access to various tools and guidelines, but managing and updating the website regularly proved to be more challenging than anticipated. Further, BeLok did not use newsletters or professional trade courses, for example, as ways of communicating results and lessons learned, and its use of public seminars was also modest. Although BeLok recognised the importance of external dissemination to other actors in the construction and real estate cluster, there did not seem to be a clear strategy and operational plan for this type of wider dissemination.

Another important but less frequently used avenue of dissemination was technology procurement and competitions. Competitions were part of the technology procurement method that the Swedish Energy Agency used to stimulate development and implementation of energy efficient methods and new techniques. This approach was intended to prove to the market actors that BeLok was an important player in the development of commercial building and demonstrated in practice the vision for future technologies. At the same time the innovative owners demonstrated their willingness to make energy efficient decisions and buy new energy efficient technologies. Although the approach was deemed successful by BeLok, a number of practical challenges for future improvements were also voiced. These included, among other things, the maturity of technologies, securing sufficient number of competitors and describing specifications and evaluation methods in advance.

Findings

The evaluation of the BeLok network pointed at the following findings:

- BeLok strongly praised the Swedish Energy Agency for its strategy of using innovative real estate owners as brokers for implementing new and improved energy technologies.

- Although the organisational structure was considered supportive for innovation through the regular and informal BeLok meetings and highly committed members, the evaluation called for additional members in order to increase impact.
- BeLok provided an excellent venue for meeting new suppliers and ideas, where they had the opportunity to gain vital experience in front-running projects and to be accepted by a strong group of buyers.
- The development activities tended to focus on technical issues rather than behavioural and management issues.
- The approach to innovation was based on pursuing 'a good idea' rather than on a systematic analysis and strategy addressing pressing needs, long-term challenges, market segments, etc.
- The evaluation strongly recommended that BeLok should intensify dissemination activities with regard to upgrading the website, launching broader types of dissemination activities such as site visits and workshops, customising information and tools for different target groups, improving the procedure on technology procurement, etc.
- Although BeLok had successfully made stepwise improvements of its operations, the evaluation pointed out a serious challenge in the innovation process, namely how to measure the overall effect of BeLok. Hence, the evaluation recommended three steps:
  - demand measurable baselines and targets for improvements in all new projects;
  - implement a new decision process as a supplement to the 'idea and belief' method; and
  - conduct a new evaluation of BeLok after another five years of operation.

Case 3: Danish network AlmenNet

The following section summarises a previous evaluation in Danish of AlmenNet by Davidsen and Bertelsen (2014). Table 13.4 provides an overview of AlmenNet.

Table 13.4 Overview of AlmenNet

	AlmenNet
Country	Denmark
Established	2004
Scope	Process improvements
Themes	Diverse set of themes
Funding organisation	Private foundation
Co-funding	Min. 50 %
Membership	Social housing companies
Secretariat	Part-time National social housing association
Activities	Demonstration projects Development of tools etc. Dissemination

Background

The background for establishing AlmenNet was a reform of the public support schemes for refurbishment of social housing via the National Building Fund (in Danish, Landsbyggefonden) and a series of separate funding applications to the large Danish private philanthropic foundation Realdania. From 2004 this led the foundation to gather its support for development and demonstration projects within the social housing sector in a new joint development programme called 'Securing the future of older social housing schemes'. The purpose of this programme was to develop and test a new model for establishing and managing innovation networks in order to strengthen the role of clients as change agents. The specific objectives of the programme were to develop, test and document:

- new tools and concepts that could support future proofing of older social housing and with potential application in housing generally;
- new and targeted development networks for clients to establish a permanent development environment among social housing organisations; and
- the new tools and concepts applied in demonstration projects.

Realdania financed the development and testing of the model in relation to a series of demonstration projects by the National Building Fund with the longer-term objective of embedding the networks in relation to the National Building Fund or a similar organisation when the support from Realdania expired.

Organisation

The collaborative efforts took place in the period 2004–13 through the following phases:

- 2004–5: Launching of the development programme and innovation network that was named AlmenNet. Emphasis was placed on establishing the website www.almennet.dk, establishing different innovation networks and launching several development projects.
- 2005–7: Focusing of the development collaboration in AlmenNet and preparation of a joint innovation model covering five areas: management; dissemination and collaboration; training and coaching; development activities and guidelines; and evaluation and testing.
- 2007–8: Establishing the independent development association AlmenNet on 20 November 2007 with its own statutes and board. This also included the transfer of the intellectual rights to the results of the programme from Realdania to the association effective from 1 January 2008.
- 2008–9: Cooperation between the development programme and the new association AlmenNet on anchoring the association broadly in the sector, completing 18 development projects from the development programme and launching 7 new additional development projects.

- 2010–13: The association AlmenNet continued development of its business with its own priorities of action, and the last development projects were completed. At this point, AlmenNet had 50 members representing some two-thirds of all 550,000 public housing dwellings.
- 2012–13: The development programme was completed with an evaluation.

In the period 2004–7 the development programme was managed by a steering committee. The members included representatives from Realdania, the National Building Fund and two social housing organisations. The steering committee was supported by a programme coordinator responsible for the daily management of the programme. The core of the development programme was a number of individual development networks within seven different subjects such as tenants' involvement, early planning and industrialisation. Initially each of these networks was composed of a number of social housing construction clients and supported by a development consultant. Later on, the development projects were managed by the housing organisations themselves in relation to actual refurbishment projects. From 2008 AlmenNet continued the development activities on its own while also taking over the responsibility for publishing the results of the demonstration projects that was initiated at the end of the development programme.

Activities and achievements

In total, 25 development projects were initiated. The total budget of the 25 development projects amounted to some EUR 2.5 million (DKK 19 million) including VAT in the period 2004–13. The first 18 projects were supported with more than EUR 1.5 million (DKK 11 million), including VAT through the development programme of Realdania, while the remaining 7 projects were supported by various funds applied for during the development programme.

Initially, emphasis was on comprehensive overall planning, prospective analysis, tenants' involvement and preparation of master plans. Later on, focus was extended to include the subsequent building phases of design, execution and delivery including tenants' involvement, project management and handover. In the recent period focus has turned towards more interdisciplinary innovation themes which cover all refurbishment phases. These included so-called 'staged' refurbishment, industrialisation, energy and environmental management, training and information and communication technology.

The development activities resulted in more than 30 different types of publication that were freely available at the homepage of AlmenNet. These included 4 pamphlets (in Danish 'AlmenHæfter'), 7 guidelines (in Danish 'AlmenVejledninger'), 4 tools (in Danish 'AlmenVærktøjer'), 5 test reports (in Danish 'AlmenAfprøvninger') and 11 documentation reports (in Danish 'AlmenRapporter'). Table 13.5 provides an overview of the different types of publications distributed on different subject areas.

Table 13.5 Types of publication distributed on subject areas

	Pamphlets	Guidelines	Tools	Tests	Reports	SUM
Information and communication technologies					4	4
Industrialisation		1			6	7
Innovation and learning					6	6
Training					7	7
Development network					3	3
Evaluation		1				1
Tenants' involvement	4	5	4	5		28
Planning and requirements	4	4	4	5	4	21
Finance and contracting	4	2	3		5	14

Source: Adapted after Davidsen and Bertelsen, 2014: 31

Findings

The evaluation by Davidsen and Bertelsen (2014) stated that the development programme and subsequent organisation AlmenNet successfully:

- established a new innovation model for social housing associations which became embedded in a permanent organisation;
- disseminated lessons learned through various means such as the establishment of a new homepage and a number of development networks on specific subject area;
- developed and disseminated a comprehensive toolbox with a wide and diverse range of guidelines, tools, etc.;
- provided professional development support and extensive knowledge exchange;
- delivered training, course activities and guidance and common principles for learning on the job; and
- assisted demonstration projects, although it proved challenging to develop and implement evaluation, benchmarking and systematic testing due to the long lead times of finalising the demonstration projects.

Despite the generally positive evaluation of the results of the development programme, Davidsen and Bertelsen (2014) point out that a long-term transformation of the sector's performance, competitiveness, etc., cannot be realised through a number of individual development projects. Instead there is a need for joint strategies that on one hand outline common long-term objectives and on the other provide space for and support to development clusters.

Analysis and discussion

The three innovation networks in Sweden and Denmark have a long history dating back to 1989, 2001 and 2004. The scope of all three networks was to drive change, but the emphasis was slightly different. Whereas the two Swedish networks seemed to emphasise procurement of technology and focused on the building professionals of the construction industry, the Danish network was less concerned with technology and more concerned with process-related issues and reaching out to other social housing clients. This was also reflected in the themes being addressed by the networks. The two Swedish networks focused on energy savings, while the Danish network addressed a diverse set of development themes.

The funding sources of the networks displayed both similarities and differences. The two Swedish networks were financed by a public agency, while the Danish network was financed by a private philanthropic foundation. All three networks were required to deliver co-funding from members and participants that as a minimum equalled the funding obtained from the public agency or foundation.

As a logical result of the different foci of the networks the membership base differed between the three networks. All three networks covered a significant share of the potential membership base. The networks were all supported by a part-time secretariat and with the secretariats closely linked to an existing interest organisation.

The activities of the networks included the same three main types: demonstration projects, development of tools, etc., and dissemination activities. The priority of each of these activities and the actual execution differed between the networks. To exemplify, BeBo excelled in well-structured and extensive use of demonstration projects, BeLok emphasised regular members' meetings, and AlmenNet adopted an elaborate approach to publication types. Table 13.6 provides a comparative overview of the three innovation networks.

After having revisited the three examples of client innovation networks as evaluated previously in their own rights, this investigation returned to the theoretical framework of transition theory and reanalysed the three cases as examples of niches being initiated by and embedded in the sociotechnical regime of construction interacting with long-term landscape developments.

Each of the three client innovation networks could be viewed as niches being established by concerned and committed clients from fairly well-defined market segments that make up part of a sociotechnical regime – residential facilities for BeBo, commercial facilities for BeLok and social housing for AlmenNet. In this regard, the niches were established by incumbent members of the prevailing sociotechnical regime rather than by outsiders or newcomers. While this provided a number of advantages, it also represented challenges. Among the advantages were the extensive knowledge by the clients of prevailing problems and solutions related to for example energy savings, the legitimacy brought by clients to ensure funding from a public authority such as the Swedish Energy Agency and ease of access to establishing demonstration projects. Among the challenges were the inherent focus on incremental solutions rather than disruptive innovations, the

Table 13.6 Overview of the three innovation networks

	BeBo	BeLok	AlmenNet
Country	Sweden	Sweden	Denmark
Established	1989	2001	2004
Scope	Technology procurement	Technology procurement	Process improvements
Themes	Energy savings	Energy savings	Diverse set of themes
Funding organisation	Public agency	Public agency	Private foundation
Co-funding	Min. 50 %	Min. 50 %	Min. 50 %
Membership	Residential property owners	Commercial facilities owners	Social housing companies
Secretariat	Part-time Client association plus consulting firm	Part-time Client association plus university	Part-time National social housing association
Activities	Demonstration projects Development of tools etc. Dissemination	Demonstration projects Development of tools etc. Dissemination	Demonstration projects Development of tools etc. Dissemination

risk of privatisation of public politics (see e.g. Pedersen *et al.*, 1992) and the pitfalls of using demonstration projects as the predominant strategy for change. Hence, the client innovation networks were facing a crucial dilemma of being closely embedded in an existing sociotechnical regime and simultaneously maintaining a critical distance to the very same regime. This leads to the first lesson with regard to niche management by clients:

> Lesson no. 1: Client innovation networks as niches require elaborate strategic capabilities in order to manage the delicate balance between proximity and distance to the existing sociotechnical regime in which construction clients are embedded.

A sociotechnical regime is composed of a number of elements such as technology, user practices and application domains (markets), symbolic meaning of technology, infrastructure, industry structure, policy and techno-scientific knowledge (Geels, 2002). The success or failure of transforming a sociotechnical regime rests not only on the ability to alter one element that is deemed important. Rather, being a system implies that one element cannot be changed without changing the others, or that changing one element may rather be the result of concerted efforts targeted at changing some other elements.

Technologies and techno-scientific knowledge are elements that were of particular attention of all three client innovation networks and which the networks were trying to transform through their different range of demonstration, development and dissemination activities. While the effort by both BeBo and BeLok was focused on reshaping the markets by stimulating the development and delivery of energy-saving services and products, and having less focus on developing clients' organisations themselves, AlmenNet was more focused on directly transforming and improving the practices of clients themselves, which in turn may have led to changes in the construction industry as a second-order effect. Other elements of a sociotechnical regime such as culture/symbolic meaning, infrastructure, industrial structure and sectoral policies were less in focus as objects for transformation by the three networks. Obviously these elements are important constituents of the sociotechnical regime, but these were largely considered as a backdrop to the innovation networks. This may, however, become problematic for the innovation networks as they may prove to be hindrances to changing the other elements. Hence, this leads to the formulation of a second lesson for client innovation networks to address:

> Lesson no. 2: Client innovation networks need to carefully examine the individual elements of a sociotechnical regime, how the different elements are interlinked and how the elements and their linkages may be obstacles or stepping stones in order to develop flexible yet robust strategies for change.

Turning the attention towards the dynamics of transition, the three cases displayed different characteristics with regard to the transition pathways being followed. The establishment of the three innovation networks as niches by construction clients as incumbent regime actors could be perceived as a response to landscape developments driven by long-term concerns of improving the built environment. The two cases of BeBo and BeLok displayed characteristics that followed the combined transition pathways of transformation and reconfiguration. Transformation appeared as landscape pressure from a long-term national and international movement towards a sustainable built environment. Reconfiguration appeared since construction clients as regime actors attempted to adopt new innovations developed by new suppliers or, in the cases of BeBo and BeLok, rather by putting pressure on existing suppliers to deliver new products and services. The case of AlmenNet displayed characteristics that are closer to a transition pathway of transformation in which the actors of the sociotechnical regime adjust to criticism. Hence, AlmenNet could be viewed as a response to long-term national concerns of being able to continue to provide affordable housing for all in an ageing building stock.

How the three client innovation networks as niches responded to different landscape developments shaped the transition paths that emerged. For example, aligning the objectives and activities of the two Swedish innovation networks to the grand societal challenge of sustainability facilitated the provision of public funding for network activities. Hence, a client innovation network needs to consciously align, oppose or take a third position in relation to landscape developments in order to improve the chances of success. As the network moves forward it will constantly need to be reflexive on the mutually counteracting, reinforcing or reproducing effects of landscape developments on the niche and vice versa. Consequently, this leads to the formulation of a third lesson for client innovation networks to address:

> Lesson no. 3: Client innovation networks need to consciously position themselves in relation to landscape developments and act reflexively on the potential effects of landscape developments on the network.

Conclusion

This chapter re-analysed how three groups of clients have formed strategic innovation networks in order to act as change agents in the construction and real estate industry. It is worth noting a characteristic difference between client innovation networks and client associations. Client associations may include innovation activities, but they are effectively trade organisations put in place to represent the interests of clients. Hence, client associations may ultimately be more concerned with maintaining the status quo rather than stimulating change. Innovation networks on the other hand will typically have a more narrow scope focusing on stimulating change in client organisations and/or the construction industry.

This analysis demonstrated how clients can act as change agents towards both client organisations and the construction and real estate industry. Client innovation networks may stimulate innovation by formulating and implementing common and ambitious demands and specifications for improved products, processes and knowledge. They can take the lead by demonstrating and documenting the value of new and better products and processes. They are also playing a role in the dissemination of best practices. However, this chapter also underlined that not all strategies and methods are equally successful and appropriate. In conclusion, the chapter pointed at three strategic dilemmas to be addressed by clients with regard to managing niches of innovation, aligning with landscape developments and embedding change in the sociotechnical regime of construction.

References

BeBo (2014). *BeBo Inriktning 2014–2015* (in Swedish: *BeBo objectives 2014–2015*). Stockholm: BeBo. Available at: http://www.bebostad.se/ (Accessed 15 May 2014).

Bertelsen, N. H. (2008). *Evaluation of BELOK 2007*. Internal document. Hørsholm: Statens Byggeforskningsinstitut, Aalborg Universitet.

Bijker, W. E. (1995). *Of Bicycles, Bakelites, and Bulbs. Toward a theory of Sociotechnical Change*. Cambridge, MA: MIT Press.

Brandon, P. and Lu, S.-L. (eds.) (2008). *Clients Driving Innovation*. Oxford and Massachusetts, MA: Wiley-Blackwell.

Brown, K. A., Hampson, K. D. and Brandon, P. (eds.) (2005). *Clients Driving Construction Innovation: Mapping the Terrain*. Brisbane: Icon.Net Pty Ltd.

Brown, K. A., Hampson, K. D. and Brandon, P. (eds.) (2006) *Clients Driving Construction Innovation: Moving Ideas Into Practice*. Brisbane: Icon.Net Pty Ltd.

Brown, K. A., Hampson, K. D., Brandon, P. and Pillay, J. (eds.) (2008). *Clients driving construction innovation: benefiting from innovation*. Brisbane: Icon.Net Pty Ltd.

Courtney, R. (2008). Enabling clients to be professional. In: Brandon, P. and Lu, S.-L. (eds.). *Clients Driving Innovation*. Chichester: Wiley-Blackwell, 33–42.

Davidsen, H. and Bertelsen, N. H. (2014). *Etablering af AlmenNet 2004–09. Udredning om forløb og resultater af Realdanias udviklingsprogramme 'Fremtidssikring af ældre almene bebyggelser'* (in Danish: *Establishment of AlmenNet 2004–09. Analysis of progress and results of Realdania's development programme 'Securing the future of older social housing schemes'*). Copenhagen: SBi, Aalborg University.

Dosi, G. (1982). Technological paradigms and technological trajectories: a suggested interpretation of the determinants and directions of technical change. *Research Policy*, Vol. 11 (3), 147–62.

Edler, J. and Georghiou, L. (2007). Public procurement and innovation: Resurrecting the demand side. *Research Policy*, Vol. 36 (7), 949–63.

European Union (2014). *Directive 2014/24/EU of the European Parliament and of the Council of 26 February 2014 on public procurement and repealing Directive 2004/18/EC*. Brussels: Official Journal of the European Union. Available at: http://eur-lex.europa.eu/legal-content/EN/TXT/PDF/?uri=CELEX:32014L0024&from=en (Accessed 8 October 2016).

Geels, F. W. (2002). Technological transitions as evolutionary reconfiguration processes: a multi-level perspective and a case-study. *Research Policy*, Vol. 31 (8), 1257–74.

Gottlieb, S. C. and Haugbølle, K. (2010). *Evaluation of Beställargruppen Bostäder (BeBo): A strategic analysis*. Hørsholm: Statens Byggeforskningsinstitut, Aalborg Universitet..

Hambrick, D. C. and Fredrickson, J. W. (2001). Are you sure you have a strategy? *The Academy of Management Executive*, Vol. 15 (4), 48–59.

Haugbølle, K. and Bertelsen, N. H. (2008). Energy savings: the client as change agent. In: Foliente, G., Lützkendorf, T., Newton, P. and Paevare, P. (eds.). *Proceedings of the 2008 World Sustainable Building Conference – SB08: Connected – viable – liveable* (Vol. 2). Melbourne: CSIRO Publishing. pp. 1595–602.

Haugbølle, K. and Boyd, D. (2013). *Clients and Users in Construction: Research Roadmap Report*. CIB Publication 371. Rotterdam: CIB.

Haugbølle, K. and Boyd, D. (2016). *Clients and Users in Construction. Research Roadmap Summary*. CIB Publication 408. Rotterdam: CIB. Available at: http://site.cibworld.nl/dl/publications/pub_408.pdf (Accessed 28 August 2016).

Haugbølle, K. and Gottlieb, S. C. (2007). From bureaucracy to value-based procurement: The client as change agent. In: Haupt, T. C. and Milford, R. (eds.). *CIB World Building Congress 2007: Construction for Development*. Full Proceedings CD, pp. 790–801. Rotterdam: CIB.

Haugbølle, K. and Vogelius, P. (2016). *Re-evaluating BeBo – the Swedish procurement group for housing: A follow-up analysis*. Copenhagen: SBi Forlag.

Haugbølle, K., Forman, M. and Bougrain, F. (2015). 'Clients shaping construction innovation'. In: Ørstavik, F., Dainty, A. R. J. and Abbott, C. (eds.). *Construction Innovation*. Chichester: Wiley Blackwell, 119–34.

Hoezen, M. (2012). *The Competitive Dialogue Procedure: Negotiations and Commitments in Inter-organisational Construction Projects. PhD Dissertation*. Enschede: University of Twente.

Högdal, K. (2013). *Halvera Mera. Slutrapport (In Swedish: Half is more campaign: Final report)*. Stockholm: BeBo. Available at: http://www.bebostad.se/wp-content/uploads/2013/11/2013_30-Halvera-Mera-slutrapport.pdf (Accessed 27 October 2013).

Hommen, L. and Rolfstam, M. (2009). Public procurement and innovation: towards a taxonomy. *Journal of Public Procurement*, Vol. 9 (1), 17–56.

Jensen, P. A., Nielsen, K. and Nielsen, S. B. (2008). *Facilities Management Best Practice in the Nordic Countries: 36 cases*. Lyngby: Centre for Facilities Management, Realdania Research.

Kemp, R., Rip, A. and Schot, J. W. (2001). Constructing transition paths through the management of niches. In: Garud, R. and Karnoe, P. (eds.). *Path Dependence and Creation*. Mahwah, NJ: Lawrence Erlbaum, 269–99.

Kemp, R., Schot, J. and Hoogma, R. (1998). Regime shifts to sustainability through processes of niche formation: the approach of strategic niche management. *Technology analysis & strategic management*, Vol. 10 (2), 175–98.

Levin, P. and Larsson, A. (2012). *Rekorderlig renovering: Demonstrationsprojekt för energieffektivisering i befintliga flerbostadshus från miljonprogramstiden. Slutrapport för Norrbacka-Sigtunahem (In Swedish: Sound renovation: Demonstration project for energy efficiency in the existing multi-dwelling buildings in the million program: Final Report for Norrbacka–Sigtunahem)*. Stockholm: BeBo. Available at: www.bebostad.se/wp-content/uploads/2012/12/RR_1-3_Norrbacka_Sigtunahem_120912.pdf (Accessed 24 February 2017).

Miller, R. (2008). Clients as innovation drivers in large engineering projects. In: Brandon, P. and Lu, S.-L. (Eds.). *Clients Driving Innovation*. Chichester: Wiley-Blackwell, 88–100.

Nelson, R. and Winter, S. (1982). *An evolutionary theory of economic growth*. Cambridge, MA: Belknap-Harvard.

Oudshoorn, N. and Pinch, T. (eds.) (2003). *How Users Matter: The Co-Construction of Users and Technologies*. Cambridge, MA: The MIT Press.

Pedersen, O. K., Andersen, N. Å., Kjær, P. and Elberg, J. (1992). *Privat politik. Projekt Forhandlingsøkonomi. (In Danish: Private Politics. Project Negotiated Economy)*. Copenhagen: Samfundslitteratur.

Rip, A. (1995). Introduction of new technology: making use of recent insights from sociolcmgy and economics of technology. *Technology Analysis & Strategic Management*, Vol. 7 (4), 417–32.

Rip, A. and Kemp, R. (with Schaeffer, G. J. and van Lente, H.) (1998). Technological change. In: Rayner, S. and Malone, E. L. (eds.). *Human Choice and Climate Change*, Vol. 2. Columbus, OH: Battelle Press, 327–99.

Wahlström, E. (2014). *Teknikupphandling av värmeåtervinningssystem i befintliga flerbostadshus: utvärdering (in Swedish: Technology procurement of heat recovery systems in multi-storey buildings: Evaluation)*. Stockholm: BeBo. Available at: www.bebostad.se/wp-content/uploads/2014/02/2012_16-BeBo_V%C3%85V_slutrapport_140120.pdf (Accessed 27 October 2014).

Postscript

Facing the changing world of clients and users

David Boyd and Kim Haugbølle

Introduction

This book has explored the changing nature of the relationship of clients and users with buildings, society and the construction industry. This was structured around three themes: agency, governance and innovation. This framework gave the book coherence against the complexity of the subject. One thing that was clear is that the world of clients and users is changing and some of this is happening quite rapidly and even chaotically. This chapter tries to draw the learning from the book into this changing world in order to explore the future. To do so it focuses on five megatrends:

- population development;
- resource depletion;
- geopolitical globalisation;
- technological disruption; and
- individual advancement.

None of these are new but their significance for the built environment both on the demand side and the delivery side has been addressed only cursorily. A new research agenda that faces this future needs to be developed.

This book has allowed us to look at the current state of research on clients and users. We understood from our work on the CIB W118 roadmap (Haugbølle and Boyd, 2013) that we are at an early stage of development of this research. The research field is moving towards establishing generalised accounts and exploring contextualisation versus the possibilities of universal theories that define the new discipline and which allow us to work in the future to strengthen our understanding of clients and users in construction. We are also seeking to establish a portfolio of case studies which demonstrate the practical use of theory. Our choice of the book structure around the client actions of agency, governance and innovation gave us a format for this universal conception and its applicability in practice. This chapter considers the extent of achievement of this desire and what we need to do next to move it on.

Chapter 1 (Bang *et al.*) considered how clients working together could have agency through client associations. However, such a cooperative approach is certainly not universal and even the UK version has a significant ideological separation to the Danish approach. Similarly, the corporate real estate model of Heywood and Russell (Chapter 2) relates to a large corporate client set within a fully developed neoliberal state where the private sector has taken over many public functions; again not a universal position. The comparison between Sweden and France presented by Bougrain and Femenías (Chapter 3) also shows many differences making universal statements difficult. The philosophical work of McAleenan and McAleenan (Chapter 4) stands out in that they see a universal morality of health and safety; however, the application of this in some locations seems very far off. The danger is that we are left with an individualistic national approach to agency which does not allow us to develop a global position.

The contributions on governance then start with a global difficulty, namely that of funding models for clients (Amidu, Chapter 5). Financial models are now becoming increasingly universal and, although many clients are adopting neoliberalism paradigms for action, there are still governments who take a profound role in financing construction activities. Thus, we know that the application of these models to the public sector, whether in Scandinavia or the UK (even under collective EU rules) is vastly different – and likely to be given even more separation as a result of Brexit. There are still large differences in the role that the public sector takes in developing buildings and their approaches to building. Translating these financial models to, for example, New Zealand, Korea, Hong Kong or Australia gives us a governance differentiation which is very complex to understand. Looking at alternative economic models, then, for example in China and smaller developing countries, becomes a challenge. These governance differences even present themselves in issues such as insurance in Denmark (Haugbølle, Chapter 6), project management capabilities in Korea (Jung and Kang, Chapter 7), the delivery of infrastructure in the Netherlands (Volker and Hoezen, Chapter 8) or of roads in Nigeria (Obunwo *et al.*, Chapter 9), showing each to be quite local affairs; yet we can see both global similarities and universal ramifications.

The expectation would be that innovation is universal because of the rapidity and ease of communications. The contributions on BIM by Widen (Chapter 12), and, in part, Thurairajah and Boyd (Chapter 11) bear this out showing software to be universal, but BIM standards and application are not. However, the procurement model of housing in Hong Kong by Fung and Yeung (Chapter 10) is innovative and also relates to the unique political, structural and social situation of this location. This leads us then to also see the study of innovation responsibilities of clients as a local act, where it has meaning in locations where the belief in markets is not universal thus it requires influential public sector clients to deliver better outcomes and these clients need to intervene in creative ways, for example with client innovation networks in Sweden and Denmark (Haugbølle *et al.*, Chapter 13) and again as regards procurement to gain improvement in Hong Kong (Fung and Yeung, Chapter 10).

We are now aware that we are seeing research reported here that is quite specific and context dependent. It is difficult to use this to give recommendations for a wider global perspective and context. We come back to our roadmap where these problems were set within a framework (shown developed in Figure 14.1 below) which identifies the central position of the client in relation to the entities (buildings and society, business and society, business and users, and the construction industry) that affect it or it serves. These entities do identify that clients are location dependent but are also changing. To move research on clients on, we need to develop the model so that it accommodates these different circumstances and changing times. This book, then, is partly doing this, but it is also heralding the need for much more work to be undertaken from a client's perspective. Clients, too, have to act effectively in a changing world and differing circumstances, whether about buildings or about their organisation. This, then, is the theme of this postscript that we need to set a new agenda for all the groups: clients, users, industry, academics and policy makers, to work together to prepare for the future.

Preparing for the future

Studies of the future involve extrapolations, projections and scenarios. Their purpose is as much to prepare people to have the capability to manage the future as for predicting the future. Thus, they need great care in their development, interpretation and use. Harty *et al.* (2007) critiqued a range of studies about the future construction industry. They identified concerns that much of the work *'reproduced, or reinforced, the current rhetoric'* and pushed certain developments and technologies as drivers of change rather than reflecting their stakeholder self-interest. However, they concluded that future studies *'can offer a way of thinking about complexity, change and an array of potential consequences and barriers'*. In order to avoid these problems for clients' future research, we have adopted a more general future study, namely the 2016 report *An OECD Horizon Scan of Megatrends and Technology Trends in the Context of Future Research* (OECD, 2016). The OECD has a good reputation in such work because its approach is reflective and reviewed by a wide audience. This was produced for the Danish Agency for Science, Technology and Innovation from existing material from a wide array of international sources and organisations with the purpose of stimulating research into the challenges, opportunities and capabilities of the future. This reports' purpose and generality, and its handling of complexity, meets our needs for critical exploration of the future for clients. We acknowledge that this presents a western centred view of the world and change herein, however, it does attempt to address a much broader international perspective.

The OECD report (OECD, 2016: 6) sees five megatrends:

- Growing, migrating and ageing: the twenty-first-century human population covering demographics, international migration and urbanisation.

- The water, energy, food and climate nexus: time for joined-up thinking covering water, energy and food security, and climate change.
- The changing geo-economic and geopolitical landscape covering globalisation, the roles of states, localisation and global power shifts.
- A moving frontier: how digitisation will drive economies and shape the ways we work covering technological change, economy, productivity and jobs, and financialisation.
- Wealth, health and knowledge: the great global divide? Covering wealth and inequality, health and well-being, access to knowledge and societal change.

This breadth of analysis is required for the study of clients in a future world as clients are not solely about building but about responding to the needs in a changing world. The world's population will continue to rise, with the greatest concentration being in Africa, while elsewhere a greater proportion of older people will emerge. This will have an impact on buildings either because more are needed or because the form needs to be developed to better respond to user needs. In addition, the ageing demographic has an impact on societies because of the smaller proportion of working people having to fiscally support this demographic development. OECD (2016: 7) speculates that 'technologies that enhance physical and cognitive capacities could allow older people to work longer, while growing automation could reduce the demand for labour'. This population is expected to be sited in bigger cities leading to stresses on utilities and the potential for slum formation and health epidemics.

To cope with this, advanced cities will manage their infrastructure in an interconnected way to become smart cities which will more efficiently manage resources. However, the growing population and its concentration puts stress on natural resources particularly water and food. Energy use will rise and the consequences of climate change will make these resources vulnerable. As a consequence, OECD (2016: 7) indicates that '[c]limate change mitigation will require ambitious greenhouse gas emission targets are set and met, implying a major shift towards a low-carbon economy by mid-century.'

The driving force of the world's economy will shift to the East with a number of new players (states, multinationals and NGOs) wielding power differentially. Globalisation, involving investments, people, goods and information, will characterise this economy but also provide problems from counter-currents producing instability and conflict. Since the provision of buildings is typically through capital-intensive projects with high levels of risk, such instability may have detrimental effects on building activity and its financing.

Businesses will further adopt digitisation with greater efficiencies from integrated design, production and supply. The digital world will still be a source of innovation and entrepreneurship based around application development enabled by open source approaches. OECD (2016: 7) predicts that 'machine learning and artificial intelligence will further disrupt labour markets, with perhaps half of total employment at high risk of becoming automated over the next two decades'. How this will impact on buildings, the industry and clients is surely an issue worth considering.

Although overall world poverty will decrease, the inequality will increase with pockets of extreme deprivation. There will be a major shift in family composition, with more single person units and couples without children. Education will still be a key aspect of advancement and this will be more available to women, hence changing the work demographic and family expectation. Although infectious diseases will be more under control, there will be a rise in antibacterial resistance. OECD (2016: 7) concludes that '*Non-communicable and neurological diseases are projected to increase sharply in line with demographic ageing and globalisation of unhealthy lifestyles.*' The unhealthy lifestyles involve food and lack of exercise which cause some non-communicable diseases (e.g. diabetes). Thus, the built environment could be a positive stimulus for better eating and more exercise, as well as accommodating people with challenged capabilities. These all impact on the variety of building types required and put greater pressures on their performance.

Although digitisation will be the most apparent technology, other developments will occur in biotechnologies, advanced materials, and energy and environment. Energy and environment are problem areas seeking technologies to overcome them, whereas the others are solutions looking for problems to solve. The report focuses on ten of these emerging technologies:

- the internet of things (IoT);
- big data analytics;
- artificial intelligence;
- neuro-technologies;
- micro and nano satellites;
- nanomaterials;
- additive manufacturing;
- advanced energy storage technologies;
- synthetic biology; and
- blockchain.

The central position of the built environment means that, even if these technologies are not directly used in its design, construction and operation, they will be associated with the client's business and so affect the form and operation of the building in ways that may be neither apparent or intended.

The report sees that there are synergies and even trade-offs between the various megatrends, thus seeking research opportunities to investigate or even exploit these. The megatrends are global, that is, it is impossible to avoid them and collectively they are often referred to as globalisation. These megatrends are all impacted on by the developments of technology. Technology is mostly presented as a solution, however, as a number of chapters in this book imply, the development and impact of technology is much more complex than merely a solution to a problem. Technology often has unintended consequences and gives rewards only to particular parties promoting them.

These future changes cannot happen without a development in the built environment, in other words the built environment is both a barrier and an enabler of society facing the future. The built environment is a barrier to change because of its long-term existence and because changes in the built environment require prior investment with little certainty that it will be successful for the new world. Against this regressive view, however, is the fact that the built environment has been fundamentally flexible in the past in accommodating the development of the present. Indeed, something about it supports this, whether it draws people together or finds space that merely allows it to happen. This has been one of the creative roles of clients and users in remaking their spaces for the new tasks at hand.

This chapter, then, reflects on this foresight and considers what research challenges present themselves for clients and users in the future. It works through the CIB W118 Roadmap framework to challenge practitioners and researchers to address the future. Figure 14.1 presents these five megatrends against the conceptual framework of clients from the W118 roadmap.

Buildings and society

More than anywhere in our conception of clients, does their multiplicity of involvement and dynamic of change come into effect than in the relations with the built environment and society. This emphasises that society regulates and

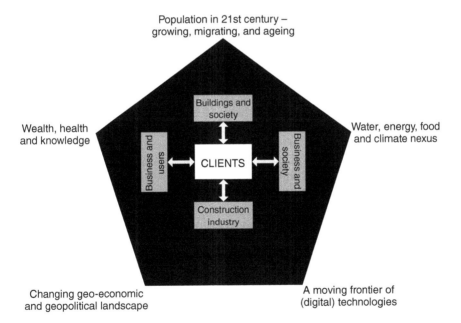

Figure 14.1 Megatrends and conceptual framework

Source: Adapted from Haugbølle and Boyd, 2013

shapes buildings around social needs; the built environment enables society to develop but, at the same time, limits its possibilities. Society also intervenes by giving grant funding for certain building types or giving favourable tax regimes for particular client activities.

The key social trend is the rise in population either internally (e.g. through ageing) or from migration. Society is going to have much more diversity and have to cope with pressures of use. This creates questions about the form of the built environment and the speed of development. Will society become segregated in megacities with gentrification in some location and slums in another? These social divides are seen around the world with gated developments and compounds where buildings are protected by security guards. While in shanty town and no-go areas, people survive with meagre resources and in constant danger. There are a growing body of texts warning about these tensions (e.g. Urry, 2016) which provide a basis for studies of buildings and society.

This then impacts on the sustainability nexus and raises questions about how infrastructures and communities can be made more resilient both to these population demands and to the climate change pressures. Climate change pressurises society to produce buildings that moderate the climate's destructive tendencies whilst not inducing further destruction because of the buildings themselves, the operation of buildings and their associated activities.

The changing geopolitical and economic order in any one location will be part of a global influence and change. Thus, there are challenges from the rise of the multinational company and its international flow of people and resources requiring a unification of identity and a standardising of product (Urry, 2016). The move to a globalised world puts us into conflict with many issues that will expose themselves in buildings or the use of buildings. Thus, we must carry out research into how buildings and society are in conflict and how the global and local tension develops new futures.

Technology has been a tremendous boon for humanity, providing it with the ability to transform the world and to reduce the uncertainties of life. A writer like e.g. Kelly (2016) sees technology's role in creating a positive future, that is, one with less problems and more capability. In the near past, our social success came from physical technologies, of which buildings are an early example. These older technologies were focused on new capabilities and the way that materials can be used effectively and efficiently. However, our most recent technological development has been around information and communication technology, which gives us many opportunities for different developments and new relationships with people. Thus, there are developing questions which challenge the need for buildings at all because of virtual association. In addition, we must look at how buildings can respond to the immediate requirements of society both from utilising big data and from our technological capabilities to make buildings more dynamically responsive.

The social shift in wealth, health and knowledge presents more social challenges but also opportunities (Castells, 2000). The need for human survival in an environmentally and socially benign world may be an aspiration that is unrealisable from our current courses of action. There is a need to fundamentally explore the long term and how we are caught in short-term thinking. Thus, the sustainable challenge for buildings and society is to develop this long-term thinking and to moderate the destruction in the short term. At one level we must consider the role of buildings in this destruction. However, climate change issues drive new needs of buildings, whether this is to enable them to respond to excessive heat, flash floods or violent storms. Our fundamental challenge, however, is in the way that we handle our existing building stock, whether we can bring it up to date or whether it is a question of new building. Some of these issues require more detailed enquiries into the role of legislation and regulation and how these impact on society.

Business and society

What occurs in buildings, that is, the business, is formed and reinforced by society, but the success of the business depends on the dynamic interaction of the business, its buildings and its context. Part of the business could be the ownership of the building and this dichotomy of interest can cause conflicts in clients. Real estate futures tend to be investigated by large consultancy firms (PWC, 2014) with a repositioning of asset management. Individual countries deliver business reports, for example CSIRO (2016), which explores business futures for Australia.

The business can be in conflict with society, and businesses operate as power systems that seek to influence society and often ask society to present regulation in its favour (Hahn *et al.*, 2010). Businesses themselves, or the public services, operating in the building, will be affected by the growth in population. This is an opportunity for businesses to have more sales but also a problem because of the pressure of space. Thus, businesses will change around the new demographics and require new or modified buildings to accommodate this.

Issues around sustainability will certainly affect clients and their businesses. The way that clients will need to modify their businesses and the way that buildings can support this new sustainable business will make useful research. Sustainability can become a business in itself and the role of the circular economy can also be facilitated by the built environment in which the business is located.

As well as the fact that populations are migrating, businesses are becoming more and more global. This globalisation trend is facilitated by new communications technologies and these strong virtual connections can mean that the business is separated from the society in which it is located. Thus, as we consider businesses and society, there is a challenge to research the nature of power of global finance in procuring and operating buildings in different locations around the world and determining whether the cultural or political barriers are negative inhibitors or provide necessary protections.

The way new technology changes the nature of business and the way these use buildings becomes a major area for research. Businesses are focusing on new technologies' ability to automate business practices. The consequences of this for society are also very challenging and must form a growing topic of sociotechnical research. Often, businesses are focusing on the ability of new technologies to automate business practice. There is a challenge, though, to understand the way that new technology changes the nature of business; and the way businesses use buildings differently becomes a major area of inquiry (ECTP, 2005). The consequences of this for society as well as for buildings are very challenging and should form a growing topic of social and architectural research.

Buildings and users

The user is becoming a more important influence in building design as clients understand that the building supports the efficiency and effectiveness of those working, studying, living, healing and undertaking leisure in it. It is not just the form and layout of the building which is important but the way that this is managed and actively utilised. The concept of building performance is becoming more recognised and being used to drive the design and construction of buildings (Mallory-Hill *et al.*, 2010). The challenge, from the changing demographics to a greater ageing population, is for the building and its operation to accommodate more elderly people and those with disabilities, for example dementia. However, issues of space allocation and standards put pressure on clients to use space more efficiently; how this can be done to support users is crucial research.

Users are also concerned about the quality of their building environment and the way it contributes to their health, comfort and wealth. The challenges then are to accommodate users, who are becoming more particular about this environment, against the trends to develop megacities which make the environment more challenging and expensive to moderate. The safety and security of this environment within a changing geo-political and economic landscape will also become more sensitive and require changes in buildings. The internal and external building environment are aspects that will be influenced by technology development, in particular the availability of big data on building performance; the way this data can be better accessed and the building controlled is likely to become a key research area. Again, the role of buildings in supporting users is fundamentally important because of the buildings' potential nurturing abilities. The research challenge then is to create a new built environment that supports people as they adapt to issues such as climate change, work–life pressures and social stressors whilst reducing resources use and the building not exacerbating the situation further.

Users are increasingly adopting a range of new technologies, in particular digital technologies that impact on the way the built environment and buildings are operated. The Internet of Things (IoT) provides the necessary foundation for smart buildings, for example remote management of temperature control of homes. Increasing use of home office working raises the requirement for the design of homes and at the same time reduced working hours at the office leads to the proliferation of new office concepts. As work is no longer a place but rather something that is done independent of space, clients, businesses and the construction industry need to develop new creative responses to these dynamic changes in work patterns, technology use, etc. Similarly, new family types such as Living Apart Together stimulate new demands and requirements with regard to smaller and more efficient homes. Improving our understanding of these developments will be of importance for developing adequate responses.

The built environment provides opportunities for wealth, health and knowledge of its users, but evidence to support or reject such claims must be assembled in a systematic manner. At one level, buildings need to be developed that bridge the poverty divide, providing accommodation which both facilitates people to develop and to feel secure. This issue is poorly researched as there are few resources or little potential for profit. Other research needs to be done on the performance demands of buildings, for example a study has already been conducted on how the design of schools can assist students in improving their learning (Barrett *et al.*, 2015). More work of this type needs to be done in hospitals and homes. Thus, the way that building environment is a stimulus to users, providing direct emotional support, encouraging action and facilitating community needs to be better understood. This becomes more important in a changing world where past norms of design may no longer be successful.

The construction industry

The supply of buildings to clients by the industry, and how they are maintained, continues to be a key subject. As part of globalisation, there is a nascent global construction industry with some companies working internationally. This reiterates the need for research on a global/local divide when it comes to work and product expectations and the way tariffs, procurement, building codes and skills licensing influence this dichotomy. There is a challenge of materials resource where, as we source materials more globally, we need to be clear about the consequences for producers, suppliers, transporters and contractors.

The construction industry has a huge impact on the environment due to its enormous use of resources, indirect impact on raw material extraction, energy consumption during the use phase, etc. But the construction industry also plays a profound constructive role in, for example, implementing climate mitigation plans. An earlier report on the future sustainability of construction (Bourdeu *et al.*, 1998) analysed the construction delivery and compared the response of 14 different countries. Others have placed studies in particular countries (Hampson and Brandon, 2004), although with global themes. Business opportunities may arise out of concerns related to developing a circular economy where the reuse, recycling and upcycling of materials can foster new businesses and may lead to the development of new building technologies. Research and development activities can help generate valuable insights into how clients can play a significant role as stimulators of such new developments through their procurement practices and the role users are playing in creating the built environment.

The technological effect on the construction industry will advance far beyond BIM and its building design and production capabilities to new forms of automation, new forms of production and new sources and analyses of data. This will provide many opportunities for research and development. This is useful for clients, as it gives added value beyond the construction phase, because digital technologies allow us to monitor components and systems continuously, the data from which can be used for more effective use and control; but, just as importantly, it will give us a long-term record of utility from which we can learn to improve the fabric and the building operation. Technology can also be used to bring together different clients and clients in different locations. At one level, this provides for the collective need of clients for information and experience about building; the strategy needs to provide a practical means of making this happen. At another level, clients need to engage collectively to demand the technologies that they need to help them. This demand-led approach to technology is less common but again the strategy needs to find a means of giving this agency to clients. However, as regards technology, we are left with many dilemmas which need researching. Although there is an inevitability about the advancement of technology, clients need to be shown that they have choices and that positive benefits of efficiency can often be offset by social problems. Thus, research on clients, who want to facilitate the industry to change, should adopt a comprehensive analysis where the technology, the building environment and society are assessed together.

The industry has a responsibility for health, wealth and knowledge improvement thus the industry may be forced to become a concern of clients. This returns us to the impact of regulation and legislation. Clearly issues of health, safety and welfare could become a major deliverable of a modern construction industry. The client can be required to have a major role in transforming the construction industry into a sustainable industry. Research then must be undertaken to show how the building industry can be made resilient and sustainable, assisted by clients, but also how it can reduce its own impact on the environment and society.

Moving forward: bringing research and practice together

Fundamental to the future of the environment of clients is the tension between the local/contextual and the global/universal and so research must try to highlight the importance of, and explore, these tensions, and possibly find some balance between them. Given these opportunities to deliver more for clients and users in the way of research and development, it is critical that a practical implementation strategy is developed. We fundamentally believe that this research must involve practitioners, indeed we need to emphasise the needs for research to be *for* and *with* clients, not just *on* clients. In that sense, we see the way that clients' agency and the governance of research needs to be supported. We are also keen to have critical studies so that the research process enables clients to learn and to appreciate the consequences of their actions. Thus, practitioners need to be part of creating the strategy and making it work for them and wider society. A future research report for construction clients and users like OECD (2016) would be helpful in order to stimulate discussion.

In addition, we need to engage with clients internationally as this helps us to explore the tension of the local and global. Although at one level there may be a convergence of clients' needs, the context of buildings, the legislative and regulatory environment and the construction industry are particular to a country. We have seen that client associations can be successful in certain locations to forge a collective spirit between clients (particularly public clients); however, we believe it would be useful and interesting to create associations between clients with similar requirements in different countries. This task requires some governance and communication protocol; in this the CIB commission W118 could be an initiator and facilitator of such research.

Apart from a futures research report on clients and users, there needs to be a call for papers to explore the identified challenges and to create future change ideas. There are opportunities for this explorative work to be co-produced between academia, clients, building professionals and policy makers. The stimulus to action of this constituency is critical and this work needs broadcasting widely. Work needs to start on this agenda immediately. This will maintain momentum yet provide space to undertake meaningful studies which can be well presented in reports and papers. This book has convinced us of the need for W118 and for initiating and promoting research into clients around the megatrends introduced here. These major issues and developing technologies fundamentally challenge our current thinking on agency, governance and innovation. The issues also impact on all of the clients' relationships, whether with society, their own business world, the people they employ or on the industry that they commission. We need to develop our understanding of clients and users so that it works for communities of both practitioners and scholars. This is no mean feat but it is essential for the research environment to advance this work. There is much work to be done but we trust that this growing community of practitioners and scholars will find a new purpose in doing it.

References

Barrett, P., Davies, F., Zhang, Y. and Barrett, L. (2015). The impact of classroom design on pupils' learning: Final results of a holistic, multi-level analysis, *Building and Environment*, Vol. 89, 118–33.

Bourdeu, L., Huovila, P., Lanting, R. and Gilham, A. (1998). *Sustainable Development and the Future of Construction*, W82. Rotterdam: CIB.

Castells, M. (2000). *The Rise of the Network Society: Economy, Society and Culture*. Oxford: Blackwell.

CSIRO (2016). *Australia 2030: Navigating our uncertain future*. Canberra: Commonwealth Scientific and Industrial Research Organisation (CSIRO).

ECTP (2005). *Challenging and Changing Europe's Built Environment: A Vision for a Sustainable and Competitive Construction Sector by 2030*. European Construction Technology Platform. Brussels: ECTP, European Commission.

Hahn, T., Kolk, A., and Winn, M. (2010) A New Future for Business? Rethinking Management Theory and Business Strategy, *Business & Society*, Vol. 49, 385–401.

Hampson, K. and Brandon, P. (2004). *Construction 2020: A Vision for Australia's Property and Construction Industry*. Brisbane: Cooperative Research Centre for Construction Innovation.

Harty, C., Goodier, C. I., Soetanto, R., Austin, S., Dainty, A. R. J. and Price A. D. F. (2007). The futures of construction: a critical review of construction future studies, *Construction Management and Economics*, Vol. 25 (5), 477–93.

Haugbølle, K. and Boyd, D. (2013). *Clients and Users in Construction: Research roadmap report*. CIB Publication 371. Rotterdam: CIB. Available at: http://site.cibworld.nl/dl/publications/pub_371.pdf (Accessed 28 August 2016).

Kelly, K. (2016). *The Inevitable: Understanding the 12 Technological Forces That Will Shape Our Future*. New York: Viking.

Mallory-Hill, S., Preiser, W. F. E. and Watson, C. G. (2010). *Enhancing Building Performance*. Oxford: Wiley-Blackwell.

OECD (2016). *An OECD Horizon Scan of megatrends and Technology Trends in the Context of Future Research*. Copenhagen: Danish Agency for Science, Technology and Innovation. Available at: http://ufm.dk/en/publications/2016/files/an-oecd-horizon-scan-of-megatrends-and-technology-trends-in-the-context-of-future-research-policy.pdf (Accessed 12 November 2016).

PWC (2014). *Real estate 2020: building the future*. London: Price Waterhouse Cooper.

Urry, J. (2016). *What is the Future?* Cambridge: Polity Press.

Index

Milton Keynes UK
Ingram Content Group UK Ltd.
UKHW040446071024
449327UK00020B/1025